AUTOMATION

FRIEND OR FOE?

AUTOMATION

FRIEND OR FOE?

by

R. H. MACMILLAN

DEPARTMENT OF ENGINEERING
UNIVERSITY OF CAMBRIDGE

With the aid of automation, I am of the opinion that we shall enter a new phase of fine living. FRANK G. WOOLLARD

CAMBRIDGE: AT THE UNIVERSITY PRESS

1956

CAMBRIDGE UNIVERSITY PRESS
Cambridge, New York, Melbourne, Madrid, Cape Town,
Singapore, São Paulo, Delhi, Tokyo, Mexico City

Cambridge University Press
The Edinburgh Building, Cambridge CB2 8RU, UK

Published in the United States of America by Cambridge University Press, New York

www.cambridge.org
Information on this title: www.cambridge.org/9781107651661

First published 1956
First paperback edition 2011

A catalogue record for this publication is available from the British Library

ISBN 978-1-107-65166-1 Paperback

CONTENTS

To

ANYA

LIST OF ILLUSTRATIONS

vii

The line-block on p. 54 is reproduced from *Metal-working Production* by permission of the McGraw-Hill Publishing Company Limited.

INTRODUCTION

> Once upon a time, a Hindu sage was granted by Heaven the ability to create clay men. When he took earth and water and fashioned little men, they lived and served him. But they grew very quickly, and when they were as large as himself, the sage wrote on their foreheads the word DEAD, and they fell to dust. One day he forgot to write the lethal word on the forehead of a full-grown servant, and when he realized his mistake the servant was too tall: his hand could no longer reach the slave's forehead. . . . This time it was the clay man that killed the sage.

Is there a warning for us today in this ancient fable? Are we in danger of being destroyed by our own creations? The perils of unrestricted 'push-button' warfare are apparent enough, but I also believe that the rapidly increasing part that automatic devices are playing in the peace-time industrial life of all civilized countries will in time influence their economic life in a way that is equally profound. At the same time, our knowledge of these mechanisms, *rightly* applied, could, it seems to me, help us to avoid some economic troubles; and it may also assist in a better understanding of many natural processes that are still obscure.

In industry the use of automatic devices enables us to make more goods more cheaply and, ultimately, with less capital outlay. In the military sphere their application makes possible the design of equipment that could not conceivably be operated otherwise.

And as they are extensively used for both purposes throughout the world, it follows that, with the relative shortage of manpower in the West, our only hope of retaining our position in the world is to install automatic equipment as fast as we can. I know these are rather sweeping assertions to make about the importance of automatic control, but I hope to justify them in this book; they are certainly endorsed by leading authorities.

For example, Major-General Appleyard, speaking as Chairman of a national conference in Margate on 'The Automatic Factory', said that British industry would have to be completely reconstructed in the next few years; and the truth of this statement was generally accepted by those present, who must be in as good a position to judge as anybody.

The basis of this reconstruction is the current movement towards a great extension in the use of automatic devices, to take over the simpler kinds of work previously done by the men in charge of machines. The word 'automation' has been applied to this process and widely adopted. It was first used several years ago by engineers of the Ford Motor Company in the United States to describe their methods for automatically conveying workpieces between successive machines; but since then it has come to have a much wider significance, implying any process in which the lower functions of a human operator—both physical and mental—are taken over by self-acting devices. Interpreted in this way, the word 'automation' clearly means more than mere mechanization, and I think the use of a new term may therefore be justified, though personally I prefer the more clearly defined expression 'automatic production'. In fact, I like to regard 'automation' as a convenient and euphonious portmanteau word, formed by telescoping the phrase above, which is rather cumbersome for constant repetition.

Automation is a new word: but does it stand for a new thing? It is doubtful if a direct answer is possible, but I shall try to give in these pages the main facts on which one might found an opinion. The book is based upon a number of talks given on the Third Programme of the B.B.C. under the titles 'Automatic Control' and 'Automatic Production'. In preparing them for publication, I have radically rearranged the material along the lines of a series of lectures given at the Royal Institution in the autumn of 1955, and have added much that had to be omitted in the first place because of the limited time.

First of all, we must be quite clear about what 'automatic control' means. A convenient way to do this is to refer, in the first chapter, to a few of the historical landmarks in the development of automatic devices. It is remarkable how long some of them have been in use; but even so, we have only recently come to appreciate that a common principle underlies their operation. This is the principle of *negative feedback*, also explained in this chapter. But control is only one component—though a vitally important one—in the process of making a fully automatic plant. We shall show that this is so by tracing, in the second chapter, the evolution of the techniques of automatic production.

The early control systems were relatively crude when compared with the devices used nowadays in industry or for military purposes. When these more sensitive controls were first made it was found that they were apt to misbehave in a most extraordinary way, plunging themselves into a state of violent oscillation. When I deal with this, in the next chapter, we shall, I hope, get some insight into the problems of control system design: naturally we shall be thinking of the basic principles, rather than of mechanical details. In doing this we shall also consider how to take account of the presence of a human operator in an otherwise automatic system.

Having got an idea of what automatic control means and of how automatic production can be achieved, we are ready to consider some of its economic advantages and also the difficulties which its use is liable to entail. Some of these difficulties are likely to be alleviated in the near future by the use of automatic calculating machines—electronic and analogue computers, as they are called—whose main characteristics are accordingly discussed in a separate chapter.

These machines are likely to have an important influence on the future developments of automation, which are considered in the last chapter. After showing there how our knowledge of feedback systems is influencing many branches of science

I shall then make an estimate of the effects that this current tide of automation is likely to have on our economy. I shall also try to guess some of the developments we may expect to witness in the near future. I realize that venturing out of one's own special field of experience in this way is hazardous, but I believe that engineers ought to be alive to the social and economic implications of the machines and processes with whose design and operation they are concerned.

Assembling and organizing this varied material has been a fascinating task and I wish to express my gratitude to Dr Archibald Clow, producer of the original broadcasts, for his willingness to sponsor a newcomer to the medium and for his great help in criticizing the first draft of the scripts: it is not perhaps generally realized how much the final script may owe to the comments of a skilful producer. I am also much indebted to my wife, who patiently listened to the scripts more times than I care to remember and made criticisms that were most useful in giving me a lay point of view. This is the view that matters, I believe, for the attitude that most of us adopt towards automation will determine whether machines are, in fact, to become our friends or our foes.

My hope is that this new branch of technology may eventually enable us to lift the curse of Adam from the shoulders of man, for machines could indeed become men's slaves rather than their masters, now that practical techniques have been devised for controlling them automatically. And in this way only can we make our factories of the future into places fit for men to do work worthy of their capacities. President Reuther, of the American Congress of Industrial Organizations, who must have studied the implications of automation as closely as any man, has reached this same conclusion and expressed it in these words: 'Economic abundance is now within our grasp if we but have the good sense to use our resources and technology, fully and effectively, within a framework of economic policies that are morally right and socially responsible.'

Cambridge, 1955

CHAPTER I

THE DEVELOPMENT OF AUTOMATIC CONTROL

A much wider use of instruments and control equipment has become an essential rather than a desirable feature of process operations.   S. W. J. WALLIS
Chief Instrument Engineer, The British Petroleum Co.

THE earliest automatic device consciously contrived is probably the pressure control invented by the Frenchman, Denis Papin, who made the first pressure cooker in 1680 by placing a heavy weight on the lid of the pan. He was thus the originator of the steam safety valve, which is one of the simplest and most widely used of all regulators today. He called his apparatus, in a contemporary translation, 'A new digester or engine for softening bones' and claimed that by its use the housewife could 'extract nourishing juices from bones which would otherwise have been abandoned as but poor prey by ye hungry dogs'. Papin is still commemorated by a handsome statue at the summit of a lofty flight of steps in his birthplace, Blois.

During the eighteenth century, various automatic regulators were applied to windmills: the fantail to turn the sails into the wind; a feathering mechanism adjusted the inclination of shutters on the sails, to control the speed; and an apparatus lifted the upper mill wheel automatically when the speed became excessive, so as to prevent overheating the grain. A very similar mechanism, called a 'centrifugal governor', was designed by James Watt in 1788 to control the speed of his steam engine. This was the first regulator to be employed extensively, and it is used, with comparatively minor modifications, on every engine and turbine today.

We also have several examples of regulators in our homes. The speed of a gramophone turntable must be held very steady, or reproduction of the records would be ruined. This is done by means of a small unit, very like Watt's governor, which provides a braking effect, as the speed rises, by pressing a felt pad against the motor shaft. A similar device controls the return speed of the automatic telephone dial. A hinged float measures the level of the cold water in the tank at the top of the house, and keeps it constant by opening the supply tap when the level falls. The same principle was once applied to regulate the flow of water to mill wheels: when the water level became too high a float operated the sluice gate, which allowed a greater quantity of water to by-pass the wheel.

With each of these controls the principle of operation is the same: any deviation of the controlled quantity from the desired value causes the controller to take action in such a way as to reduce the deviation. To obtain the most accurate regulation, the better controllers take progressively more action as

PLATE I

(*a*) REFINERY CONTROL DESK

In some modern chemical plants a control desk replaces the control room. The operator makes adjustments to the settings on the various automatic controllers after studying the state of the plant on the graphic panel in front of him. In the next stage of development the operator may be replaced by an electronic computer.

(*b*) WATT'S GOVERNOR

The 'centrifugal governor' designed by James Watt to control the speed of his steam engines. The spindle is driven by the engine, so that if its speed rises the balls are flung outward and thus operate a linkage which reduces the steam supply.

PLATE II

KELLER COPYING MACHINE

This machine, introduced in 1921, is typical of the automatic machinery whose use is a first stage towards the automatic factory. It is shown here shaping a die for pressing motor-car bodies; the contour follower on the surface of the model above guides the cutting tool on the work below.

the deviation increases, but for many purposes this is a needlessly costly refinement.

We have considered regulators for controlling steam pressure, water level and speed. But any physical quantity can be controlled, provided there is a means of measuring it and a means of adjusting it. One of the most important of these quantities is temperature. Devices for its control are used on the oven, the water heater and the refrigerator. The name 'thermostat' originated with Andrew Ure of Glasgow, who was granted a patent in 1830 entitled 'An apparatus for regulating temperature in vaporization, distillation and other processes'. The crucial points in the design of a thermostat are the device to detect the temperature under control and the method by which it effects control. Ure used the bending of a bimetallic strip, which depends upon the different rates of expansion of metals on heating: the same property is used in the 'rod and tube' type of thermostat frequently employed on gas-heated appliances, and first introduced about 1900. Another type of temperature detector is a metallic bellows filled with a fluid which expands on heating; its use nowadays in refrigerator thermostats was pioneered in 1925 in Britain.

Yet another household thermostat will serve to illustrate an important distinction between two basically different sorts of regulation. The control of a central heating system might work by using a measurement of the temperature of the house to adjust the fuel supply or boiler draught. This is a

closed sequence of control because the temperature
of the house affects the heat supply, which in its
turn affects the temperature. On the other hand,
the thermostat might use a measurement of the
conditions outside the house, increasing the fuel
supply when the weather gets colder or windier.
This is an *open sequence* of control: the weather
affects the heat supply, but the heat does not affect
the weather.

The success of open sequence control depends
upon just the right properties being built into the
controller. If the heating control takes account
only of the weather, the house will get cold when a
window is left open or if the thermostat is not in
perfect adjustment; but the closed sequence system
will continue to work satisfactorily in spite of such
disturbances, because the heat supply will be increased until the house does reach the temperature
desired. A closed sequence control incorporates
feedback; that is, the results of its own actions are
fed back to the regulator and modify its further
behaviour.

The controls we have considered so far have all
been *regulators*. Another important group are
used for *position control*. The earliest of these were
steam operated and were used to provide greater
forces than a man could apply unaided. The hydraulic valve, which is the essential feature of these
early position controls, is said to have originated
with young Humphrey Potter. In 1713, he was in
charge of one of the steam pumping engines in-

vented by Thomas Newcomen. Potter's job was to open and close, at the right moments, the valves that admitted and exhausted the steam. He noticed that admission was always needed when the piston was in one position and exhaust when it was in another, and this led him to the idea of making a link between the piston and the valves to get his job done automatically.

This pleasing tale is almost certainly a myth, for it is extremely doubtful if a *Humphrey* Potter even existed, although there were others of that name connected with early steam engines. But though strict historians must class the incident with Watt's kettle, Newton's apple and Washington's little hatchet, it is worth recalling for the principle it illustrates. Developments of this valve mechanism have been used in every steam-engine since, and it was also adapted for use on the steering motors designed for the early steamships. These motors, which assisted the helmsman to move the large rudders, were the first major development of automatic control after Watt's governor.

In 1868 Brunel's vessel, the *Great Eastern*, was fitted with such a motor, the movement of the rudder being controlled remotely by the position of the helmsman's wheel. The force to turn the rudder was provided by steam pressure in a cylinder, with entry of the steam controlled by a valve mechanism derived from Potter's. The novelty and ingenuity of the apparatus consisted in the fact that there was a further device, called a 'follow-up' linkage, which caused the steam valve to be progressively closed by the motion of the rudder. As a result, by the time it had reached the position indicated by the wheel, the valve was again fully closed; so no further steam was admitted to the cylinder, and the rudder accordingly came to rest in the desired position. This 'follow-up' mechanism is another instance of negative feedback, an indication of the actual position of the rudder being fed back to the desired position and subtracted from it, to control the subsequent motion.

It is possible to regard all automatic controls as different sorts of closed sequence regulators; a regulator being a control which acts so as to reduce the difference between the actual value of the controlled condition and its desired value. Just as the thermostat reduces the difference, or deviation, in temperature, so the servo-motor acts to reduce the positional deviation. With the positional controller, the negative feedback may take the form of an actual physical follow-up linkage; but though the negative feedback action is less apparent in the thermostat, it is none the less there.

A feedback linkage was also used by the Frenchman, Joseph Farcot, in his invention for powered steering. Describing this in 1872, he wrote: 'We thought it necessary to give a new and characteristic name to this novel mechanism and have called it a servo or enslaved motor.' Farcot was thus undoubtedly the originator of the name *servo-motor*, which is now commonly used for a powered position controller. In his book, *Le Servo-moteur ou Moteur-Asservi*, he defined it more precisely. A free translation is as follows: 'Any motor at the absolute command of an operator, whose hand acts directly or indirectly upon the control member of the motor, which moves so that the two go, stop, proceed and reverse together, the motor following at each step the operator's finger, imitating like a slave his every gesture.'

We have seen that the earliest servo-motors used steam for their motive power, because it was conveniently available. But compressed air or oil under high pressure can also be used. A. B. Brown of Rosebank, Edinburgh, patented the first of these—a hydraulic steering device—in 1870. Designers of large or high-speed modern aircraft are faced with just the same problem of providing sufficient force to move the rudder and other control surfaces, and in recent years hydraulic servo-motors of much the same type have been used to assist the aircraft pilot.

Farcot also proposed a *two-stage* servo-motor for the steering of large ships; and guided missiles must contain similar servo-motors, usually with more than one stage, to transform the minute electrical signals, received by radio, into the considerable forces needed to move the control fins.

Military needs have been one of the main stimulants to research in servo-mechanisms, and here one of the most important milestones is Robert Whitehead's torpedo depth control, a very advanced

design for its time. An anaeroid capsule detected
the water pressure, and thence the depth to which
the torpedo was submerged; the movements of this
capsule operated the fins, which caused the torpedo
to climb or dive, as might be required. But the
fins were not controlled solely by the capsule: in
1869 a pendulum was added and so arranged as to
measure the inclination of the torpedo; the movements of this pendulum were also coupled to the
fin control, so causing a marked improvement in
performance. The inclination of the torpedo is a
measure of the rate at which its depth is changing and
we shall see in the third chapter why one would
expect a knowledge of this to be helpful; here we
have the first use of an important principle that is
frequently applied in modern automatic controls.

In its original form, the torpedo fins were moved
by a direct mechanical linkage from the capsule and
pendulum. Later a servo-motor, operated by compressed air, was incorporated in the control. To
understand why this complication was necessary
we must remember that a servo-motor acts as a
power amplifier, because, though its movements
follow its input, the force behind them is greatly
increased.

This capacity for power amplification is one of the
most useful properties of servo-motors, for not
only can they be used to provide controlled power
far greater than could be obtained manually, but
they also make it possible to use minute, flypower
signals. This is often necessary in remote control

applications, as it is not usually convenient to transmit large amounts of power. Power amplification also permits the detection of signals without disturbing unduly the delicate instruments that produce them.

The pneumatic servo-motor in the torpedo depth control increased the force available to move the fins, and reduced the disturbing reaction on the control—the sensitive capsule and pendulum. A servo-motor was also needed to reduce undesirable reactions in the ship's gyro-compass, because any disturbance of the gyroscope would have completely ruined its direction-finding property. The presence of the servo-motor also made it possible to have repeating compasses at various stations in the ship, operated by a single master.

If it is left undisturbed, the axis of a gyroscope remains fixed in direction, a property used by Sperry as the basis of the first automatic pilot for aircraft. He designed this only a year or two after Blériot first flew the Channel. Movements of the aircraft relative to the gyro reference direction were detected with the least possible disturbance and the control surfaces operated through servo-motors. Similar principles are applied in modern automatic pilots, except that it is common practice now to measure, as with Whitehead's depth control, not only the aircraft's altitude, but also its rate of change. The radio controlled 'Queen Bee' of the 1930's and the V 1 are the most striking examples of pilotless aircraft, flown by automatic controls.

More recently there has been reported the automatic take-off, flight across the Atlantic and landing of an aircraft. Such a project involves many separate control systems.

In recent years the engineers at Rosebank have pioneered the development of the Denny-Brown stabilizer for ships. This consists of a pair of electro-hydraulically controlled fins that project from the side of the ship and oppose its rolling motion, which is detected by gyroscopes; the action is the same as that of the ailerons of an aircraft under automatic control. This highly successful device reduces the roll to negligible proportions and, though designed to assist naval gunnery, it is now being fitted to many passenger liners.

Until twenty or so years ago most servo-motors were operated hydraulically or mechanically, but since then many systems have had electric servo-motors and electronic amplifiers, both of which sometimes offer considerable advantages. Electronic amplification makes it possible to use signals of extremely low power, such as those available from radio and radar. In addition the transmission of electrical signals is simple and they are easily modified, by special circuits, to improve the performance of the control system.

Every day we benefit from the automatic volume control on our radio sets. This is an electric circuit which measures the strength of the incoming signal received by the aerial, and prevents fading when the signal decreases, by automatically increasing the amplification in the set. Those fortunate enough to have an automatic picture brightness control on their television sets are served by a similar, though more elaborate, control.

Many useful measuring instruments, such as the photo-electric cell, have electrical outputs and it is natural to make the remainder of the system electrical, especially as electric motors are usually much more easily designed and manufactured than hydraulic ones. Compared with a hydraulic motor, however, an electric one of equal power is bulky and slow in response. Furthermore, electrical energy is more difficult to store than compressed air; so that mechanical control systems are unlikely to be entirely superseded.

The most extensive and elaborate application of electrical control has been to guide the motion of naval and anti-aircraft guns. When these guns are fired they must point ahead of the target, towards the position it is expected to occupy when the shell reaches it: this is achieved by comparing, all the time, the actual gun direction with the target's future direction, as predicted by an electronic computer; when the two differ, suitable corrective signals are sent to the servo-motors that move the gun. The great accuracy attainable nowadays by naval surface gunnery is exemplified by the engagement in 1941 of the battleships *Bismarck* and *Hood*, which took place in heavy seas at the extreme range of about seven miles. The first salvo, fired by the *Hood*, actually straddled the *Bismarck*, though no shell found its mark, and a shell from the return salvo of the *Bismarck* sank the *Hood*.

The relative simplicity of electronic design has greatly increased the attractiveness of automatic controls in industry, where they are now used for regulating numerous physical quantities such as voltage, speed, tension, temperature and acidity. Provided that methods can be found for determining quickly, accurately and continuously the present state of the condition to be regulated, electronic means can readily be used to effect control.

But difficulty may arise from the accuracy of measurement demanded. A decade ago, an error of one part in a hundred might have been admissible, whereas one part in a thousand may be the current requirement; an increase in accuracy of this order is liable to demand an entirely new technique of measurement. In a servo-mechanism for ruling diffraction gratings, for example, an accuracy of a millionth of an inch is needed, and this has been achieved by utilizing the interference of light waves, whose lengths are of this order of magnitude.

We shall see in the next chapter that the essential feature of modern production technique is that the product is kept moving through the plant, as it is subjected to successive treatments. This substitution of a continuous flow for separate processes has been developed to perfection in modern chemical plant, as in the production of plastics or petrol. Such plant demands the most elaborate control. Until about 1920, the method used in the chemical industry was for a man to watch a gauge registering the temperature, pressure or flow at some point, and when it deviated from the desired value he would make an adjustment, perhaps to the heat supply to the process. He would then wait to see the effect of his first adjustment and later make further corrections as necessary. The closeness of control was dependent on the operator's skill and, at best, the method was slow. The situation is quite different now. The reading from the gauge on the plant is transmitted to a central control room, where a pneumatic device called a 'process controller' receives this information and computes what action should be taken. This decision is then transmitted continuously to the servo-motor on the plant, whose task it is to act upon these instructions. Because of fire hazards, present-day computing and signal transmission is usually done by compressed air, but flame-proof electronic controls offer advantages which will probably make them supersede the air-operated ones.

The adjustment of process controllers to get the best results is often critical and this is one aspect of automatic control we have not touched upon so far, but it is of the utmost importance. The question is the size of the deviation that will produce any definite amount of corrective action: the more sensitive the control, the smaller the deviation needed to do this. It is often advantageous for the control to be made as sensitive as possible, as residual errors, caused by mechanical friction, for example, are thereby reduced, and also the response to disturbances becomes more rapid: the system is, in fact, under tighter control.

Provided the necessary power is available, there is no objection to increasing the sensitivity of an open sequence control. But with a closed sequence system there is a serious drawback, for this reason. When an attempt is made to increase the sensitivity of a perfectly well-behaved system, a stage is reached at which it suddenly begins to oscillate violently. This is because ever-increasing signals can circulate around the closed control loop, much as a puppy chasing its own tail will go faster and faster, until it can go no more quickly. In the third chapter we shall examine in more detail when and how these self-excited oscillations can occur, and shall see what can be done to prevent them.

CHAPTER II

THE EVOLUTION OF
AUTOMATIC PRODUCTION

Flow production involves a marriage of management and mechanism, with management as the
dominant partner.   SIR LEONARD LORD
Chairman, British Motor Corporation

THE ultimate objective of industrial control is the
fully automatic factory. To achieve this, the
operation of individual machines must be made
automatic, and they must be automatically linked.
If a hand-operated machine is to be made fully
automatic, all the functions previously fulfilled by
the operator must be taken over by various mechanical and electrical devices.

Let us start, then, by considering what a machine
operator in fact does: it is surprising how much it
may amount to, even when he has an apparently
simple job, such as feeding material to a machine
which automatically performs an operation like
drilling one or more holes. In this case, he must
pick up the piece to be drilled from the waiting pile,
insert it in the correct position in the machine
and fix it: he then switches on the machine and
when it has completed its cycle of work, he removes
the drilled piece and places it with the other finished
pieces.

In addition to these obvious duties, he examines

the piece visually before and after drilling, to ensure
that it is not cracked or otherwise damaged; he
checks continually that the machine is operating
correctly, by listening to the sounds it emits and
by seeing that the drill is neither broken nor in
need of sharpening; he ensures that the pile of
waiting material is not likely to become exhausted
nor the finished pile outrun its storage space, and
he may keep a record of the number of pieces drilled.
The chips of metal drilled out must be removed
from machine and work-piece and returned to
salvage; and, finally, the operator must take appropriate corrective action, should any be required:
this may involve passing information to earlier or
later stages in the manufacturing process.

We can see now that in any manufacturing
process—whether the product is a vacuum cleaner,
a drum of oil, a box of chocolates or a packet of
cigarettes—the functions to be performed automatically can be included in one of three broad
categories. Firstly there is *material handling*, which
involves transfer, storage, loading, fixing and
unloading; also included in this category is *assembly*
and packaging. Secondly, there is *inspection* of
product and machine, both during the process of
manufacture and after it is completed: this involves
making measurements of size, state and composition
and also assessing such elusive qualities as appear-
ance and flavour. The third essential component
of automatic production is *control*; in this category
is included controlling the operation of machine

or process, the control of stock, and taking corrective action in the light of information received from the various inspecting and measuring devices.

Let us consider how each of these main functions can be made automatic. To discuss modern techniques for handling material, it is convenient first to trace very briefly the stages which have led to our present state of industrial development. In 1798 Eli Whitney first made mass production possible by manufacturing components to a sufficient accuracy to allow of interchangeable assembly. (One recalls the delight of James Watt, at this time, when it was found possible to fit a piston into a cylinder so closely that no more than 'a worn shilling' could be inserted between them.) For the next century, factories were so organized that the part being manufactured was moved to different regions in the workshop in accordance with the operation to be performed upon it. But when flow, or line, production was introduced into the motor industry, the part was made to progress instead on a continuous line, while the machines needed to perform each operation on it were brought to the line and placed along it. This was a bold departure from established practice, for it meant using many more machines than before; but the continuity of operation enormously reduced and simplified the handling of material.

Once line production was in use, it was a natural step to cut out the manual labour of transferring the material and loading the machines. The first transfer machine, making cylinder blocks, was introduced at Morris Motors in 1923; the row of work-pieces was shifted from station to station, like the guests at the Mad Tea Party, and clamped in position by hand. In the following year, James Archdale manufactured for Morris the first machine with *automatic* transfer and clamping. These early British designs preceded by some twenty years the automatic transfer machines which have since been introduced in the United States, in Britain and on the Continent. Most of the technological problems involved in the design of transfer machines have now been solved, though no doubt (as their use becomes more general) we can expect them to become simpler, more reliable and more versatile.

It is interesting to compare with transfer machine development the evolution of a machine such as the lathe, used to manufacture an article which demands a succession of turning operations. Originally, the lathe would be set up for the first operation and, when this had been completed on a number of pieces, it would be reset for the next operation and the pieces would again be inserted for further treatment, and so on. The next stage was to have a separate lathe set up permanently for each operation, and move the pieces along the line. Handling was further reduced by introducing the turret or capstan lathe, in which a number of different cutting tools can be brought to bear successively on a single work-piece. The tool movements can

PLATE III

AUTOMATIC PRESS LINE

This line of five presses arranged in a circle is controlled by one operator. The work-pieces are held throughout on the central spider arms and fed from machine to machine as the spider rotates. Advantages of the circular arrangement for transfer lines are the convenience of transfer and the fact that a single station can be used for loading and unloading, which may lead to an economy in space and labour. With transfer presses, the operating hazards normally associated with press operations are practically eliminated.

PLATE IV

TYPICAL TRANSFER MACHINE

This twenty-one-station transfer machine processes the intake manifolds of automobile engines. Tests are automatically made for leakage and air flow, and a mechanical memory enables the machine to reject faulty parts. The extreme cleanliness of some modern factories is reminiscent of a hospital.

be made automatic and, when this is done, it may be possible to perform several cutting operations at the same time. In this final stage of development—the multi-spindle automatic lathe, as it is called—instead of bringing a succession of cutting tools to the work, the tools stay in the same place and the work is presented to each in turn. There are usually six spindles, which means there are five working positions and one for loading and unloading; at each station several simultaneous operations can be performed.

The problem of assembling complicated components automatically is much more difficult than that of mechanized handling: often, the only way in which it can be done at all is by a basic redesign of the product, with automatic manufacture in view. The most striking example of this is in the assembly of radio sets, whose miscellaneous mass of electrical connexions could only be soldered automatically by an absurdly complicated machine of immense bulk.

The solution was found by J. A. Sargrove, who devised the printed circuit technique for making electronic equipment. The required circuit is made in the form of one or more layers of flat plastic plate upon which lines of metallic paint form the conductors and other simple components. The circuit is then assembled by inserting the valves and remaining components in the plates: a recent machine to do this can assemble up to twenty-four components into a printed circuit plate at the rate of 1200 an hour; so that in continuous operation this one machine could produce no fewer than ten million radio receivers in a year.

Clearly, it is useless to have machines of such immense productive capacity unless it is possible to check *immediately* that their products are satisfactory; otherwise they might turn out large quantities of defective material before a fault could be rectified. In Sargrove's circuit-making equipment, careful attention is paid to this aspect; in fact, he is himself a pioneer in the field of automatic inspection, and he has shown how the equipment can be made to go a stage further than merely accepting or rejecting the product, by enabling it to observe trends and initiate the action necessary to ensure that many fewer rejects occur, just as a skilled operator might do.

Since an essential feature of automatic production is that the operation must be continuous, so also should be the inspection processes; but the continuous measurement involved may present severe problems. Usually indirect methods must be used, as for example in finding the thickness of steel strip, moving at a thousand or more feet a minute while it is rolled out. The vibration is too great to permit direct measurement, but we can find instead both the pressure between the rollers and their distance apart, and deduce the thickness from them.

To solve these tough problems of continuous measurement, some of the latest scientific techniques are being brought out of the laboratory and put to work in factories. This has meant a fresh approach to the design of delicate apparatus, to ensure that it will still remain reliable when it is roughly handled. To gauge the thickness of a moving strip of paper or fabric, a light source may be placed on one side and on the other a photo-electric cell, to detect the amount of light passing through the strip and so indicate its thickness. For an opaque material, like steel strip, a similar arrangement with a source of radiation and a counter can be used. In the same way the density to which the tobacco is packed in cigarettes can be found by mixing with the tobacco a small amount of radio-active material, with a short life, and measuring the amount of radiation emitted from the cigarette roll. One can also find the thickness of the film on a metal surface, as in tin plating, by measuring the amount of radiation reflected from the surface; and reflected sound waves of very high frequency are similarly used for crack detection.

Often it is necessary to measure continuously the composition of a stream of gas or liquid. The carbon dioxide content of a gas stream, for instance, can be found by using the fact that the rate at which a hot wire cools depends upon the constitution of the gas surrounding it. Stream composition can also be found by measuring continuously such properties as density, acidity and viscosity, and instruments for these purposes are now being developed.

This brings us to the third component of automatic production—control. Stream properties such as temperature and pressure are easy to measure, provided that very great accuracy is not essential, and we saw in the last chapter how the deviations of the measured values from those desired are used to control the process itself. In the same way, when the measured thickness of a rolled steel strip exceeds the desired value, a servo-mechanism automatically brings the rollers closer together. This type of industrial automatic control has been intensively studied during the last ten years or so, and it is now known how to make equipment that is both accurate and reliable.

In some chemical factories, we have now reached the stage at which improved control can best be obtained by designing the plant for its *controllability*. To get the best plant dynamics, detailed attention must be given both to layout and to the construction of individual units. Recent research along these lines has drawn attention to gaps in our fundamental knowledge of Chemical Engineering.

Provision must also be made for controlling the sequence of movements of machines and handling equipment, so that successive operations are carried out in the right order, and to ensure, on transfer machines, that all the interconnected units have completed their work before the whole line is moved on a stage. At present electrical relays, that work interlocking circuits, do this. Another type of machine control, which it is more difficult to accomplish satisfactorily, is needed to determine accurately the relative motions of the cutting tool and the work-piece, so as to remove just the right amount of metal in exactly the right places and as rapidly as possible.

This problem first arose at the end of the eighteenth century when the Yorkshireman Joseph Bramah collaborated with Henry Maudslay in producing the screw-cutting lathe. In this machine the cutting tool is automatically moved steadily along the face of the material which is being turned. Bramah was a prolific inventor and we also owe to him the hydraulic press and lock, which bear his name, in addition to the beer pull, the water closet and the modern form of hydraulic valve.

In 1818 Thomas Blanchard developed the copying lathe, a machine to cut irregular shapes by copying a metal model. All the force to move the cutting tool was furnished by pressure against the model itself, but it was costly to make a mechanism strong enough to move the tool accurately, and the model soon became worn. These disadvantages are overcome in machines like the Keller copying machine, introduced in 1921. A cutting tool, driven by a hydraulic servo-motor, is constrained to repeat every movement of a detecting finger which follows a path along the surface of a wooden model, bearing quite lightly upon it. By using a rather similar electrical device, to follow a line traced on a sheet of paper, it is possible to dispense with the costly business of model-making, and

manufacture directly from the drawings, on which
the exact position of the line is detected by a photo-
electric cell.

Even the work in the drawing office can be cut
out if the machine is controlled by punched cards
or magnetic tape. A familiar example is the
punched roll which determines the tune on a
player piano. The use of punched cards can be
traced back to 1801 when the Frenchman, Jacquard,
introduced a loom which would automatically
weave any desired pattern according to the cards
with which it was fed. And the Monotype machine,
for setting up print, works in much the same way.
We shall discuss this in more detail in chapter v.

The first industry to be extensively mechanized
was food manufacture, and it still uses by far the
greatest proportion of automatic equipment. In
1794, Oliver Evans erected in Philadelphia a continuous flour mill, in which the grain was untouched
by hand throughout the process of grinding and
bagging for dispatch. In 1833, biscuit making
for the Royal Navy was mechanized, and the
continuous monorail for pork packing was introduced in Chicago in 1869. Though designed for
the dismemberment of hogs, this was the fore-
runner of today's assembly line, which made
possible the mass production of motor cars, when
it was first introduced by Henry Ford.

One of the earliest fully automatic lines was
installed as long ago as 1930, in Milwaukee, to
make 10,000 chassis a day for cars, a rate which

could not have been approached with a conventional plant of anything like the same size. There were nine separate units linked by automatic handling devices; these units inspected the incoming strips of steel, rejecting all that were unsatisfactory; they then cleaned, bent and drilled the remainder, assembled and riveted them, washed and painted them, and finally dispatched them to store.

In 1947, Sargrove set up his fully automatic Electronic Circuit Making Equipment, to make radio sets with printed circuits, and this has been claimed as the first 'automatic factory'. Whether the claim is justified or not depends upon the definition adopted for an automatic factory, but undoubtedly it was a most remarkable achievement.

In several countries, parts of motor cars are made without human intervention, but it is not yet economic to make assembly automatic—though drastic redesign might make a big difference here. Since 1951, for example, a fully automatic plant, manufacturing aluminium pistons for lorry engines, has been operating in Russia: chunks of metal are fed to the furnaces at one end of the line and pistons, packed ready for shipment, emerge at the other, untouched by hand throughout the process of manufacture. Even the removal of waste metal from the machines is automatic. An interesting feature of this plant is the fact that the line is adaptable to make several sizes of piston, according to the demand, and changeover from one to the other involves no delay: this feat would be most difficult to accomplish in a manually operated line.

Many of the processes in oil refining, in paper making and printing, and in the manufacture of plastics are nowadays automatic, the operators' functions being purely supervisory. The same applies to much of the processing and manufacture of food and confectionery, though some of the inspection processes have so far defeated our ingenuity; and many ordnance factories have automatic lines for making bombs and ammunition. A more recent achievement is the 'ribbon machine', which makes the glass envelopes of electric light bulbs and radio valves: two of these machines, occupying only a few thousand square feet of floor space, are installed near Doncaster; they can produce the complete requirements of the United Kingdom and still allow an ample margin for export. The sand and other raw materials enter the machines at one end and, at the other, finished envelopes, packed in cartons, are loaded directly into railway wagons, all the manufacturing processes being entirely automatic throughout.

We shall see in the final chapter what new fields of manufacture automation is likely to invade in the next decade or so. It is certain that current achievements are but the van of the great army to follow.

CHAPTER III

CONTROL SYSTEM
DESIGN PROBLEMS

The characteristics of automatic controllers [for chemical plant] are well known, and their actions can be matched with a plant and process to give the best possible control with given equipment.

J. MCMILLAN
Central Instrument Laboratory, Imperial Chemical Industries

THE earlier control devices, which we considered in the first chapter, were designed on a basis of previous experience coupled to trial and error: and so long as one is content with controls that are relatively inaccurate and slow in response, no great problems of design arise. But as soon as we wish to improve the performance or even go a stage further and obtain the best possible performance in given circumstances, we must look at the underlying theory—the mathematics—of these devices. To explain the nature and results of this theoretical work, it is convenient to take as an example a thermostatically controlled central heating system, as its familiar behaviour aptly illustrates general principles.

We have already seen that the first step towards improving the performance of a control system is to use a *closed sequence* to compensate for the effect of random disturbances. This means that the controller measures the actual value of the controlled quantity, which might be the temperature of a house, and compares it with the desired value, which has been set on the controller. The sequence of operation here is closed, since the error determines the controller action, which in its turn reduces the error. The amount of action taken for a particular size of error is called the *sensitivity*. As the sensitivity is increased the system responds more rapidly and usually controls more closely; but there is a limit to the sensitivity that it is possible to use.

This was first recognized over a hundred years ago by the Astronomer Royal, Sir George Airy. Airy was concerned with the design of speed regulators for his telescopes, to give them a rotation exactly equal and opposite to that of the earth, so that they would point in a fixed direction. He devised several mechanisms for this purpose, but in every case he found that as he increased the sensitivity and reduced friction between the parts, to secure greater accuracy, a stage was reached at which the control did not operate smoothly at all; the telescope motion became uneven, periodically accelerating and decelerating with increasing violence. This form of oscillation has since been called *hunting*.

The question naturally arose: was the tendency to hunt when the accuracy was increased a characteristic of *all* regulators? It was desirable to know, in addition, whether there was, in fact, a practical upper limit to the accuracy attainable and, if so, what it might be. There is, of course, a profound difference between the mere recognition that closed sequence controls are liable to hunt and the formulation of a theory that will give precise answers to questions such as this. Airy's own analysis of his telescope controls was published as long ago as 1840. He showed that hunting might be expected to occur if the natural period of vibration of the regulating mechanism was closely related to its period of rotation: this led him to propose modifications which were found, when tried, to improve their performance sufficiently for his purpose.

The next major contribution came from Clerk Maxwell, the genius who predicted the existence of wireless waves before they had been detected physically. This great applied mathematician was also the father of automatic control theory, for the methods of analysis which he introduced were very much more general in conception than those of Airy, and were in fact the only ones available until twenty or so years ago.

It is not surprising that Maxwell made such an outstanding contribution to Control Theory when one recalls his extraordinary abilities. His first scientific paper was published when he was fifteen years old and he was elected Professor at London University when he was only twenty-four. He decided to go into retirement after ten years there, and it was during this period that he made his researches on speed regulators, but six years later he returned to Cambridge as the first Professor of

Physics at the new Cavendish Laboratory and it was then that he did his celebrated work on electromagnetism.

In order to appreciate Maxwell's work, we must first discuss in more detail the types of behaviour to be expected from a closed sequence control system. Let us consider, for example, a central heating system with a thermostat, that automatically cuts down the heat supplied by the boiler when the temperature of the house exceeds the desired value.

What happens if, when conditions are steady, we wish to raise the temperature of the house by, say, 5°? The knob indicating desired temperature is moved to the new value, and the thermostat thereupon takes action appropriate to an error in temperature 5° low; but naturally the house does not get hotter immediately, as the boiler must first raise the temperature of the water circulating in the radiators. When the house eventually does reach the desired temperature, it will still go on getting hotter, because of the excess heat that is by then stored in the radiators. It follows, then, that the temperature of the house overshoots the desired value; and a similar argument shows that, on cooling down again, it undershoots it. The cycle is then repeated.

Probably the sizes of the overshoots will progressively diminish, until the house finally settles to a fixed temperature, and conditions are again steady: such a system is said to be *stable*. But if the sensitivity of the thermostat is increased enough, an increasing temperature fluctuation may occur, and the system is *unstable*. When the sensitivity of the control is very low, on the other hand, there may be no overshooting and oscillation at all. The particular type of response thus depends, not so much on the nature of the control system (provided it is error actuated), as on its sensitivity.

Now, it is often important to know not only about the immediate consequence of a disturbance, but also about the state of affairs achieved later, when conditions have again become steady. When the central heating system is operating at a steady temperature, the heat supplied to the radiators from the boiler is just equal to the heat lost from the house, through the walls and roof. But what happens if the outside temperature drops, or it gets windy? The house now loses heat faster, so there must be a greater supply of heat from the boiler to attain a steady state again. But the boiler heat supply is only increased when the thermostat receives an error signal, which means that the temperature of the house must be below its desired value. It follows that when it gets cooler outside, the temperature of the house must inevitably drop to some extent, though very much less than it would in the absence of any control. All simple regulators must behave like this: a lasting disturbance causes a persistent error that is large enough to produce a control action sufficient to compensate for the disturbance. So it is evident that the greater is the sensitivity of the controller the smaller is the persistent error caused by a particular disturbance.

A small drop in temperature would not usually matter in a dwelling-house, but if our thermostat were controlling the temperature of a process in an oil refinery, it might be highly objectionable, for several reasons: a few degrees difference in temperature might seriously lower the grade of petrol produced by the plant or, even worse, a small rise in temperature (during hot weather) might speed up the reaction enough for it to get out of hand and cause an explosion.

When it is important for permanent errors to be eliminated, two methods, which are closely related theoretically, can be used. One way is to make the controller take account not only of the present temperature error but also of the cumulative error: any persistent error causes this cumulative value to mount steadily, so that the thermostat takes progressively more and more action until the error is eliminated. This type of control is sometimes called *reset action*, as it is equivalent to automatically resetting the desired temperature to a higher value when there is a tendency for the temperature to be persistently low—which is the way a human operator would achieve the same end.

Alternatively, the controller may be made so that it alters the heat supply continuously at a rate which depends on the temperature error. The heat supply is thus increased steadily until any persistent drop in temperature has been eliminated. The disadvantage both of rate control and of reset action is that, for the same rate of response, they make the control system much more liable to overshoot. That is, they have an unstabilizing influence.

There are two ways in which it is possible to counteract this tendency to instability, and they were both suggested in principle by Maxwell in 1868. In this classic paper, he first discussed the behaviour of speed regulators such as those used by Airy, and then considered those with rate control. He mentioned the existence of the unstable motions we have considered, and remarked that they were generally disregarded by the inventors of these devices, 'who', as he put it, 'naturally confine their attention to the way in which they are designed to act'! Maxwell then suggested two remedies for this unstable behaviour. One of them was to reduce the tendency to oscillate by introducing frictional forces opposing the oscillatory motion—a method still frequently used.

The other method he proposed is equivalent to introducing *anticipation* into the controller action. This idea is very simple in principle: it will be recalled that in the central heating system we considered earlier there is a tendency to overshoot the desired temperature when the control is too sensitive. This tendency can be reduced by arranging for the thermostat to take less action when the temperature error is diminishing and to take more when it is increasing. This is very like anticipation or prediction, as the controller is always acting now upon its estimate of what the error will be a short time ahead.

With refinements, this is the most commonly used means for achieving stability in control systems today. It involves a measurement, not only of the error, but also of its *rate of change*. It will be remembered that it was the function of the pendulum in Whitehead's torpedo depth control to make just this measurement. In the central heating system, a simple way of obtaining an estimate of the rate of change of error would be to measure the temperature of the water circulating in the radiators, since the difference between this temperature and that of the house determines the rate at which the house gets hotter.

It would also be a good plan to use a measurement of the temperature outside to control the boiler, so introducing an open sequence feature into this closed sequence system. If this is done, there will be a very much more rapid response to changes in the weather, for compensation for a drop in temperature is then started immediately, instead of only occurring after the temperature of the house itself has first fallen.

We have now found that to obtain an ideal central heating system, the controller must take into account three separate temperatures; firstly, the house temperature, to close the control loop and eliminate permanent errors; secondly, the radiator temperature, to give anticipation; and thirdly, the outside temperature, to give immediate warning of disturbance and accelerate the response. It will be a matter for computation and experiment to decide the best proportions in which to use these data to operate the furnace draught control.

To calculate just how much anticipation is needed (that is, how far ahead the control should work in any particular instance) and to find how great the sensitivity should be, to obtain the most rapid and accurate system which is still stable, it is necessary to write down equations, as Maxwell did. He found their solutions in some of the simpler cases, but the general problem of stating the conditions for any equation to represent a stable motion was only solved ten years later by another Cambridge mathematician, E. J. Routh.

Curiously enough, Routh and Maxwell were undergraduates together and sat for the Mathematical Tripos examination at the same time. An amusing anecdote is told of this occasion. Apparently Maxwell was so confident of his triumph in the examination that he did not even trouble to rise earlier in the morning than usual, to hear the lists of successful candidates read out in the Senate House, but sent his servant instead. On his return, Maxwell is said to have inquired of him, 'Well, tell me who's second!', and was somewhat taken aback to receive the reply, 'You are, sir!', for Routh had defeated him. For his work on

4

stability Routh was awarded the Adams Prize, an honour which Maxwell had previously won for his study of Saturn's rings.

In order to write down the equations which represent the behaviour of a control system, it is very convenient to think of the sequence of actions in terms of what is called a *flow diagram*: this can be pictured as rather like a necklace, with the beads representing the components of the system (such as the boiler, the radiators, the house and the thermostat), while the string represents the operative signals (which, in this case, are the various temperatures) that are passed from one component to the next. The necklace can either be left open or clasped into a closed loop, to correspond with the two possible forms of control.

To make use of this flow diagram idea, one must have some means of describing precisely the contribution of each component to the behaviour of the system as a whole. This can be done by developing the concept of a *time lag*, which arose earlier, in connexion with the central heating system, as the interval between the action taken by the controller and the resulting response in the temperature of the house. Every component of the system contributes to this overall time lag, which is the sum of their individual time lags. The time lag of each component is a property which does not alter with the type of signal being applied to it. It may be measured experimentally or, sometimes, calculated from the physical properties of the component.

Suppose now, that instead of considering the effects of applying a single and prolonged disturbance to a control system, we study what happens when it is subjected to a regularly fluctuating signal. As the period of time taken to complete one cycle of disturbance is made less, the time lags will probably not alter greatly, so that they will come to occupy an increasing proportion of a cycle. When they add up to half a cycle, the signals fed back round the loop tend to reinforce at all times the original disturbance and, if the sensitivity is then sufficiently high for the signal which has travelled round the loop to be as strong as the disturbance which originated it, a self-sustained oscillation can arise.

This approach, which was the one introduced by Nyquist twenty years ago in the United States, has two advantages. It provides a practical test procedure for designers and it gives an immediate insight into the factors likely to improve performance. To do this, we must first ensure that all the time lags are as small as is possible, and also reduce the total lag by using anticipation. As we have already found for the central heating system, this may involve a subsidiary feedback of information to the controller from some point other than the output.

As far as testing is concerned, the method used is to take each part of the system in turn, apply fluctuating signals to it, and measure its response. When the designer is armed with the frequency response, as this is called, of every component in his system—it is just as important, of course, to know about the process under control as about the control apparatus—mathematical methods have been devised to enable him to choose the best values for the sensitivity and any other factors which may be at his disposal.

Making frequency response tests can be a very costly business when it is necessary to put large units of equipment out of production for the purpose: finding the characteristics of a typical water turbine driving an electric generator in a plant of the Tennessee Valley Authority, for instance, cost $2000 for each hour of the test. Even so, the results were well worth the expense of obtaining them. Sometimes it is possible to make tests while the system is in normal operation. The largest test of this kind ever undertaken must be that on the Swedish electric grid, covering about 2½ million square miles, whose output (in the region of 5 million horse-power) was varied cyclically with periods between 5 and 10 seconds. The time lag was found to be about 6 seconds.

In many systems an important link in the chain of control is a human operator. To analyse the system it is just as necessary to describe his behaviour mathematically as that of any of the other components. Regarded in this light, the input to the human operator is the stimulus, received usually by the eye, and the output is the muscular reaction, which may be a movement or pressure of the hand or foot.

Experiments indicate that the component giving the closest approximation to a human response consists of two sorts of time lag, together with a degree of prediction, the amount of anticipation increasing with training and skill, just as one would expect. But such experiments are liable to give inconsistent results, on account of the variability of human reaction, depending as it does on fatigue and other extraneous influences such as anxiety. In addition, the response to a large stimulus may differ considerably from that caused by a small one, which means that a human being does not behave as a *linear* component. Let us see what is meant by this term.

All the controls discussed have been tacitly assumed to be linear, which means that the output from every component increases progressively as its input is increased; if the size of the input is doubled, so is that of the output. The most important characteristic which follows from this assumption of linearity is that when we know the response of the system to a pair of separate disturbances, we can deduce what the response would be if the two disturbances occurred together. This very useful property does not apply to a *non-linear* system, and all real systems are to some extent non-linear.

While there is only *one* sort of linearity, there are a host of different sorts of non-linearity, so we must limit ourselves to a few typical examples. One of the commonest is *saturation* of one of the components. This means that when the input signal to the component becomes sufficiently large, the output ceases to increase further. When a train is full, for instance, the fact that there are still passengers waiting on the platform will not enable it to carry any more. An important consequence of saturation is that the amplitude of oscillation of an unstable system does not increase indefinitely, but mounts to some steady value and then gets no greater. Another non-linear component is a measuring instrument whose sensitivity varies in different parts of its range. It might be good at measuring medium-sized quantities, but less good at measuring very small or very large ones: a foot-rule is excellent for measuring a length of a few inches, but is hopeless if we wish to measure tenths of an inch or many feet.

A further example of a non-linear component is the type of controller to be found on an electric iron: there is no gradation in the corrective action. If the temperature of the iron is too low, full power is switched on, and when it is too high the supply is cut off altogether. Unlike the noble Duke of York in the nursery rhyme, the power is never half-way up—it is always up or down. With many 'on-off' controls of this sort the only steady state possible is a continuous oscillation of small amplitude. 'On-off' controls are very simple and can therefore be made both cheap and reliable;

but an important reason why they are hardly used at all where high performance is demanded is the absence of satisfactory methods of mathematical analysis and design.

It is a difficulty that is by no means confined to 'on-off' control: though the theory of linear systems is for the most part worked out, non-linear theory is still in its infancy. It is the most challenging of the current problems of automatic control and one which is receiving world-wide attention, including that of a group at Cambridge University. It is an extraordinarily important subject, as there is absolutely no reason for supposing that linear systems are the most economical either in power or in first cost. It is also likely that a better performance can be obtained from a suitable non-linear control than from any linear one. Even apart from this, the designer may be forced to use a non-linear system, because of the nature of the components available or of the method of error measurement that must be used.

An example of this is a *sampling* system. The controller here receives only occasional, instead of continuous, information about the state of the error. On receiving an error signal, it takes appropriate corrective action, which lasts unchanged until a further signal enables it to make an adjustment. Interest is added to the theoretical analysis of sampling systems because it seems possible that a human being may also behave intermittently in a manner that is rather similar.

But, as yet, the most important conclusion of research into the behaviour of a human operator is that the best results will be obtained when his job is made as simple as possible, any prediction that may be necessary being performed automatically. The golden rule for the designer of a system involving a human link turns out to be 'Suit the machine to the man, rather than attempting, by elaborate training, to fit the man for a needlessly difficult task'.

It is not surprising that human reactions are so complex, quite apart from their non-linearity, when one takes into account the amazingly complicated system of interrelated control loops from which the nervous system is formed. Analysis is difficult enough when only two automatic control systems are similarly linked. This happens, for example, when we wish to control both the humidity of the air in a house and its temperature. We cannot treat this as two separate problems, for making the air moist may also alter its temperature, while making it warm must change its humidity. Theory indicates that it is possible for a *multiple loop* system of this sort to be unstable, even though each control would work satisfactorily with the other out of action. It can also happen that a multiple loop system that is normally quite stable becomes unstable when one of the variables is fixed, by external interference.

An important practical example of a multiple loop occurs in the control of a gas turbine or jet engine. The speed of rotation must be limited, or the rotor would fly to pieces, and the gas temperature must not be too high, or the material would wear out very quickly. Nor must the speed or temperature be too low to maintain the power output. But these two quantities are each dependent in a complicated way both on the fuel supply and on the size of the intake or exhaust passages, which may be made variable. This is indeed a difficult problem for the control engineer.

But even with a single loop control, the designer usually has, in practice, only a rather rough idea of the various time lags, so that he cannot expect to lay out a complete new scheme from theory alone. Analysis is none the less important, for several reasons: his calculations enable the designer to decide which factors it is most profitable to adjust to improve the performance and, furthermore, he can obtain from his equations some indication of the best performance which is attainable. Theory can accordingly provide him not only with a guide for his early designs, but also with a goal at which to aim. It may thus help him to estimate what degree of refinement is economically desirable, a matter we shall consider in the next chapter.

THE ECONOMICS OF
AUTOMATION

This is a time of transition . . . from the ancient
problem of sharing scarcity to the modern problem
of distributing abundance.   ADLAI STEVENSON
                                        7 December 1955

Selling automation to existing supervisory staff is
often more difficult than convincing the shop-floor
worker that it is desirable.   F. G. WOOLLARD

I am confident that the problems [attendant on
automation] can be solved, but only if they are
more widely known.   H. NICHOLAS
                        Transport and General Workers' Union

WHEN he was addressing a meeting of the Trades
Union Congress, which was devoted to a discussion
of automation, the representative of the General
Council said that 'the ignorance shown on the
rostrum has probably been equalled only by that
displayed on the platform'. Even allowing for
rhetoric, this remark stresses how important it is
for us to examine next why a man may be replaced
by a machine. What are its main advantages?
These seem to fall into two groups: the machine
may be doing a job that a man could not do at all,
or it may do a job that a man could perform, but he
would do it less effectively.

In the first category we might include the mechanisms that assist a man's limited strength, in such

devices as ship's steering. Automatic controls are also indispensable where it would be dangerous for a man to go; an example is into the harmful radiation near an atomic pile. Another use for automatic control is where there would be no space for a man, as in many rocket missiles, whose existence would be inconceivable without their automatic guidance and control mechanisms.

Automatic devices may be used, too, because they can act far more quickly than any man could: this makes them necessary for safety devices of all sorts, and for the control of rapid chemical reactions. These process controllers also perform a certain amount of computation. The capacity to compute rapidly and accurately is essential to the control of a highly unstable craft, such as the 'Flying Bedstead', and it is only the automatic devices through which it is controlled that enable this strange vehicle to take the air at all. The increased speed of automatic inspection means that every article can be checked instead of only a proportion: this again may lead to an improvement in the quality of the product.

A machine may control more effectively than a man because of its superior efficiency, precision, economy or reliability. The aircraft automatic pilot, for instance, replaces a man because it is more efficient: it does not suffer from fatigue, and it gives the passengers a smoother ride than the human pilot can. Because of their ability to control accurately and to within close limits, automatic controls may improve the efficiency of industrial plant. It is often necessary to maintain an elevated temperature in large storage tanks, containing several thousand tons of oil, and great savings can be effected in the steam needed for heating if the temperature is controlled automatically. In electric generating stations, also, automatic control of boiler pressure can reduce the fuel consumption considerably.

Again, owing to the more uniform flow in a chemical plant under automatic control, it is possible to operate with less storage capacity; this in turn reduces the time lags and increases the speed of operation of the plant. It is true to say, in fact, that even if armies of labour were available, no modern oil refinery could be operated manually at anywhere near its present capacity.

The automatic devices that displace men because they do the same job more cheaply are becoming of increasing importance as the cost of labour rises. A typical example here is the automatic traffic light, which replaces the policeman on point duty, who is released for more useful and less tedious work. The many thousands of automatically controlled crossings in this country effect an annual saving of several million pounds. Simple automatic devices can be made extraordinarily reliable, especially if duplication and various counterchecks on their operation are used; this makes them ideal for such functions as automatic signalling and train control, where they can eliminate errors due to the 'human element'.

We have now seen the main advantages to be derived from the use of individual automatic controls. But what is to be gained from installing fully automatic production lines? Are they going to help us towards what has been called 'the age of harmony, leisure and plenty'?

Above all, the elimination of unskilled manual tasks considerably reduces direct labour costs, a reduction that must be reflected in the price of the finished article, particularly where labour is expensive. Fifty years ago 15,000 man-hours were needed to produce a far simpler motor car than that which today needs only 1000 hours of direct labour. On their automatic piston plant, the Russians claim that direct labour has been reduced by three-quarters and unskilled work has been practically eliminated.

But, even apart from this saving, several other factors tend to make automatic production cheaper. The proportion of time during which machines are in use can be increased by 30 per cent or more and the amount of material lying idle in store and in process can be greatly reduced; this means that the productive capacity of plant is increased relative to the space occupied and to the capital equipment involved. In addition, transfer machines made on the unit principle, described later in this chapter, may actually cost less than the machines which they replace, since they can be built up from simple and standard components.

The use of transfer machinery may also effect

unexpected savings, as when transfer presses are used to form articles by pressing flat metal sheet between a series of dies. This has revolutionized manufacture from certain materials which become

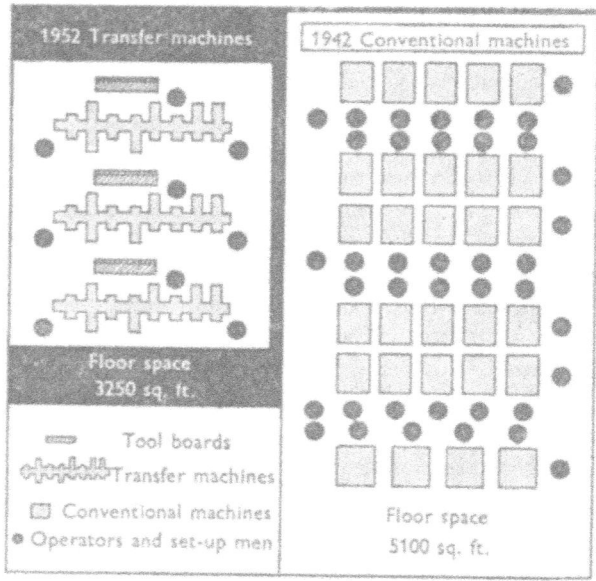

Transfer and conventional machines compared. This diagram shows the saving in both floor-space and direct labour which is effected by replacing the twenty-nine machines previously needed by three eight-station transfer machines.

brittle when they cool and must then be slowly heated to make them sufficiently flexible for further treatment, a process which had to be carried out between each pressing operation when manhandling was used. But with a transfer press one operation follows so swiftly upon another that there is no time for cooling and the intermediate heat treatment can be omitted. As a result, a piece which would previously have taken minutes, or even hours, to make can now be completed in a matter of seconds.

But the advantages of automatic production are by no means solely financial. The individual worker also gains: an automatic factory is a much safer and pleasanter place to work in and the jobs are more interesting. Human capacity is wasted on pushing and lifting, or even on the simpler control functions. This costly commodity is better employed on work that demands more intelligence—such as maintenance, supervision and planning. For the workers in automatic production there will certainly be a general up-grading of skills; and this is very desirable provided the additional skill and training are adequately rewarded.

One of the chief objections to work on a manual production line, as satirized by Charlie Chaplin in *Modern Times*, is the inevitable 'pacing' of the operative's work by the conveyor or by the time cycle of the machine; this common cause of mental distress is eliminated when such functions can be made automatic. Perhaps, as Sir George Thomson recently suggested, it may be possible to train monkeys to do the simple jobs that it is not economic to make automatic. Henry Ford evidently spoke more wisely than he can have realized when he said that the ideal factory worker would be a trained ape!

In view of these many advantages to be expected from the use of automatic devices, it is not surprising that in recent years production has increased out of proportion to capital outlay, while there has been a great reduction in working hours, together with the elimination of many monotonous and laborious manual tasks. But, in spite of this, it has been suggested that, even on new plant being erected an increase of no less than tenfold in the expenditure on automatic control would be possible, using only standard equipment now available. This estimate is based on the conjecture that, taken over industry as a whole, about 5 per cent of the capital cost would be devoted to the instrumentation, in plant fully equipped with current automatic devices. Whereas, in fact, only about $\frac{1}{2}$ per cent of British industrial capital expenditure is accounted for in this way.

Why then are not even more automatic devices in use today? There seem to be several reasons, the most important of them being expense, inflexibility of function and labour problems. Let us take each in turn.

The first cost of automatic equipment is itself bound to be heavy. And it rises steeply with complication: a machine which is expected to be versatile in function will be more expensive than special purpose equipment. Here lies the greatest advantage which man still has over any automatic control yet devised: his astonishing adaptability.

The manufacturer is on the horns of a dilemma:

PLATE V

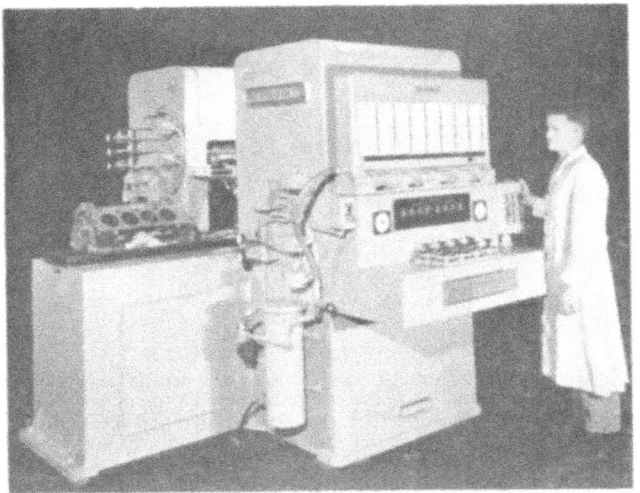

AUTOMATIC GAUGING MACHINE

Here is an example of an air-operated high-speed inspection device. It measures eight dimensions simultaneously and grades the product (a V-8 cylinder block) into one of fifteen classes. The actual sizes of the dimensions are shown on the screen at the operator's eye-level.

PLATE VI

(*a*) TAPE CONTROLLED MACHINE

The movement of the workpiece under this drill is controlled by magnetic tape fed to the computer behind. High accuracy is achieved by means of a system employing a diffraction grating to translate the movements into a series of electrical pulses. (This is a purely experimental machine; the system has now been applied to three-dimensional milling.)

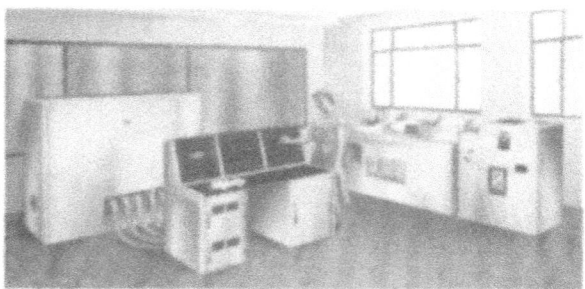

(*b*) COMMERCIAL ELECTRONIC COMPUTER

This is the latest version of HEC. The operator works at the desk in the centre. The information is fed in at the part on the right-hand side, on which the final answers are also typed, and the computation is done on the left.

he must use automatic equipment, because he cannot otherwise produce as cheaply as his competitor, who has it; on the other hand, if his competitor should bring out an improved product, which is stealing the market, he may, on account of his automatic machinery, find himself saddled with a severe loss; for workers can be transferred, but overheads on inactive machinery cannot. To run an automatic plant economically, moreover, it is essential to operate the machinery continuously; this means that production increases can only be obtained by adding to plant, and more shift-working may be necessary.

One way of reducing the functional inflexibility of transfer machines is called the 'package' technique: standard units for machining and handling are interconnected, so as to procure the particular sequence of operations the designer wants. An alternative approach is the original system of automation at Ford's, mentioned in the introduction. The workpiece is moved between successive machines on a shuttle, that travels either forwards or sideways along conveyors, while turntable and roll-over devices present it to the automatic loading mechanisms in any desired position. In this way, a number of normal machines can be linked to form an automatic line.

Another problem with transfer machines in continuous operation results from the time consumed in changing the cutting tools that need resharpening, since the whole series of connected units must be stopped to change a single tool. For this reason, a large machine might be out of action for as much as half the time. This difficulty has been overcome by breaking down large machines into several sections, which can be stopped separately while the remainder stay in action. Provision must be made, of course, for the work-pieces to accumulate in front of each section while it is stopped.

The third drawback to the use of automatic equipment is the antagonism caused by the labour shifts involved. Several skilled workmen may be displaced by the introduction of automatic machines, which can be operated by a single unskilled man. But we must remember that these machines demanded skilled design, and still require skilled attention to maintain them in good working order. The situation is clearly complicated, so let us look at this labour problem in a different way.

If capital investment were to increase while the need for manpower dropped, the consequent rise in capital's share of the national income would cause widespread unemployment and have an adverse effect on the standard of living. But the capital needed for unit output has not, in fact, increased for the last forty years, and we may even expect automation to reduce it; it follows that labour should retain at least its present share of the national income, while working much shorter hours to earn it. But this does not imply that unskilled labour will always be needed to the same extent. On the contrary, we must expect a general raising of the level of skill demanded, with the result that the living standard of the man determined to continue at the same job will inevitably drop compared with that of his fellows who have acquired additional skill.

It is absolutely essential in an automatic plant for management to maintain good labour relations, for strikes are even more disruptive than in plant that is less highly organized. The chief problem is with the men whose skills are made obsolete by the installation of new equipment. They must be trained, before this happens, for alternative work that is more interesting and better paid. The attitude of the Trade Unions, who have a monopoly of labour in a sellers' market, will be important: at present they are adopting in Britain a policy towards automatic production of 'wait and see'; they are investigating what it is likely to involve and are carefully watching such developments abroad as the guaranteed annual wage—Britain has, of course, had something very like this for some years now in the docks.

In the United States it is realized that wages must be sufficiently high to maintain the purchasing power of the workers and this is made a primary consideration in all wage bargaining. Walter Reuther has said categorically, in this connexion, 'It is my belief, and that of my colleagues in the C.I.O., that we must initiate effective action to guarantee *a constant increase* in mass purchasing power'.

Trade Unions must accept the necessity for a continual upgrading of their membership, and they should all follow the example of those which already demand technical qualifications before accepting membership, and provide, or encourage, educational facilities. In some ways, the C.I.O. structure, with an all-embracing union for each industry, appears to be more suitable than that of the American Federation of Labor or of the British Unions to cope with the labour shifts which automation must bring, and to avoid the rivalries between unions which they might otherwise engender.

Many thoughtful Trade Unionists appreciate that automatic production should benefit labour in the long run, though few would be prepared to go as far as Mr J. Edgar, of the United Patternmakers' Association, who described automation as the 'key with which we can leave this dim vault and walk out onto the upland pastures—the Elysian fields'. Most Unionists would, perhaps, be more willing to agree with their colleague who remarked recently that if the coming years do indeed turn out to be ones of leisure and plenty for the working man, then there will certainly be industrial harmony as well.

CHAPTER V

AUTOMATIC COMPUTERS

'Can you do Addition?' the White Queen asked.
What's one and one and one and one and one and
one and one and one and one?'
'I don't know,' said Alice. 'I lost count.' 'She
can't do Addition,' the Red Queen interrupted.
'Can you do Subtraction?'   LEWIS CARROLL

The four rules of arithmetic may be regarded as
the complete equipment of a mathematician.
                    JAMES CLERK MAXWELL

ONE very important class of automatic machines
has been mentioned so far only in passing—these
are the automatic calculating machines or computers. It is interesting to study them in greater
detail, as it has been realized recently that in their
latest developments they are something far more
momentous than merely convenient time savers
for mathematician and accountant: it is now clear
that in the near future they will play a vital part
in the control of automatic plant.

Calculating machines are, in fact, closely anala-
gous in purpose and in operation to factories, for
their function is the processing of information,
just as that of a factory is to process metal, food or
chemicals. The machine is supplied with certain
data which it manipulates according to the instructions it is given and so produces the information
desired.

Computers can be divided into two main classes according to the way in which they operate: firstly, there are those which are set up as a physical analogue of the problem to be solved and are called *analogue computers*. The simplest of these is the slide rule, in which lengths along the scale correspond to the numbers being manipulated: the action is not, of course, automatic. The clock, whose mechanism automatically ensures that the hour hand revolves just twenty-four times as fast as the earth, might be regarded as an analogue machine for computing the divisions of the day.

Process controllers are also analogue computers. They convert the data regarding the process deviation into a physical quantity, such as an air pressure or electrical voltage, and manipulate it in pneumatic or electric circuits to produce a measure of the control action needed. The physical processes in such a controller are analogous to the solution of a differential equation. To solve more elaborate differential equations an automatic computer called a *differential analyser* is used. Though a design was first formulated by Lord Kelvin in 1876, one of the first of these machines actually to be constructed was made at Manchester in 1935 out of Meccano parts. Such computers are very suitable for solving certain types of problems, and many others have been built since then.

A large differential analyser can be set up to solve a particular problem much more quickly when the connexions between the units are made electrical and can be completed by merely dialling the appropriate numbers, as in a telephone system. Such a computer, designed by Bush, a pioneer in this field, has operated for some years at the Massachusetts Institute of Technology. Another innovation is to effect control by feeding instructions to the machine on punched tape. One of the largest analogue machines, called *Tridac*, is now nearing completion at the Royal Aircraft Establishment. A monster like this costs hundreds of thousands of pounds to build but is worth it, if it will provide information unobtainable in any other way. Electrical analogue computers are usually called *simulators*.

The second main class of computers are much more important from our point of view. They are called *digital computers*, because they work directly with numbers or digits. The simplest and oldest digital computer (after man's own fingers and toes) is the abacus, but the idea of constructing a mechanism to perform automatically the operations of ordinary arithmetic originated with Pascal at the age of nineteen in 1642, when he made a machine to add and subtract. In 1673, Leibniz extended this machine, enabling it to multiply and divide. A little earlier Leibniz had written: 'It is unworthy of gentlemen to lose hours like slaves in the labour of calculation, which could safely be relegated to anybody else if machines were used.' After this, many mechanical calculators were developed, and one was put on the market commercially in 1820, but it was not reliable. The first of the modern type of desk calculating machine was produced by a Swedish engineer in 1891, and desk machines have since become indispensable for much scientific and clerical work. They are all, of course, hand operated, though in modern ones it is no longer necessary to 'turn the handle'. The speed of computation which can be achieved with their aid far exceeds that attainable with pencil and paper.

The first automatic digital machine was designed by Charles Babbage and largely constructed by 1829, when work on it was abandoned. The previous year Babbage had been elected Professor of Mathematics at Cambridge, although he had no degree, and he held this office for eleven years without delivering a single lecture. Babbage was both a genius and an eccentric in the grand manner, with his antipathy for barrel organs and anxiety to reform the Royal Society, 'to rescue it from contempt in our own country, and ridicule in others'. He wrote voluminously on an extraordinary variety of topics, including the economics of mass production and the development of machine tools; he showed that the cost of collecting, franking and delivering letters greatly exceeds that of transporting them—an observation which led eventually to the introduction of the penny post.

Babbage's first machine, the Difference Engine, was a special purpose computer to be used for calculating mathematical tables. In 1833 he conceived the further idea of a universal computer, which he called an Analytical Engine: this machine was to contain every essential ingredient of a large modern digital computer. To make such a machine from entirely mechanical components would today tax us to the utmost: a hundred years ago the mechanical difficulties defeated even Babbage's ingenuity.

The first universal automatic computer actually built was completed at Harvard in 1944 and called the Harvard Mark I Computer. This machine contains mechanical counters, but uses electromagnetic clutches and relay circuits for interconnexion of the units and control of the operation. It is only about one tenth of the size of the machine Babbage had conceived. In the Harvard Mark II machine, completed in 1947, all operations are performed by electromagnetic relays, of which there are some 13,000; this greatly increases the speed of operation. In one second it will perform five additions or one multiplication of two ten-figure numbers, the results being typed out or punched on tape.

Other relay-operated machines have been constructed, but the next major advance was the substitution of electronic circuits for electromagnetics. The first machine to use these was the ENIAC (Electronic Numerical Integrator and Calculator) built ten years ago in Philadelphia. It was this machine which caught the public imagination and was hailed as the first 'electronic brain', a term which is harmless provided we do not attempt to take the metaphor literally. The machine lacks many of the characteristics foreseen by Babbage and is extremely cumbersome for what it will do, as is to be expected with all early developments; with its 18,000 valves and 100 kilowatts of power consumption, it is probably physically the largest digital computer that will ever be built. It will perform 5000 additions or 40 multiplications of two ten-figure numbers in a second—a great advance on the Harvard machines.

The ENIAC and most earlier machines count in tens, but it was shown, by von Neumann and others at Princeton, that it is more efficient to count in twos and this is what all the really high-speed electronic computers do nowadays, even though it necessitates an awkward double conversion between the scale of two and the scale of ten, in which the data is known and the solutions are needed. The first computer to take full advantage of the possibilities of electronic techniques was the EDSAC (Electronic Delay-Storage Automatic Computer) completed at Cambridge University in 1949. Although it contains only 3000 valves, addition is six times as fast as with ENIAC and the machine is far more versatile.

A more recent large-scale digital computer, constructed at the Massachusetts Institute of Technology, is called *Whirlwind*, on account of its very high speed of operation. It has 6000 valves and will perform 200,000 additions or 25,000 multiplications in a second. It will do at least as much calculating in a minute as a man with a desk machine could do in a year: this immense increase in the speed of numerical computation renders entirely practical calculations which were previously quite impossibly long. Another very high-speed machine has been built at Princeton where it is affectionately known as the MANIAC. This computer is designed for 'parallel' operation, which means that when making an addition all the figures are added simultaneously, as on a desk machine: the EDSAC and most others use 'series' operation, in which the figures are taken one at a time, as in calculations with pencil and paper. Parallel machines are inherently the faster, but they are more elaborate and difficult to design.

The digital computers we have considered up to now are all extremely costly and are generally far too bulky to be suitable for commercial purposes. But before we discuss the machines of this sort which are now coming on to the market, it will be valuable to consider in rather more detail, though without going into any technicalities, how these automatic computers work. They are all basically just adding machines but, unlike Alice, they work so quickly that they do not lose count, even when adding at a rate of a million units a second. Multiplication is done by means of successive additions, as in a desk machine, and subtraction either by a reversal of the adding process or by adding the negative of the number to be subtracted. Division can be effected by a process of repeated subtraction, though other methods are often faster. Any numerical mathematical problem (and many problems in logic) can be put into such a form that the solution is obtainable from these four processes, which are performed in what is called the 'arithmetic unit' of the computer.

Like a desk machine, to which it corresponds, the arithmetic unit must have registers for storing both the numbers to be processed and the solutions found. And, just as with the material stored in marked bins in a factory, each number stored in a computer must be in a location that can be easily identified, when the number is next needed, by what is called its 'address'. The speed of operation of a computer depends largely upon the time required for access to any particular number in storage. Normally a high-speed electronic store is much more limited in capacity than one with a relatively low access rate, such as is used for storing the sort of data a man at a desk would have in a book of tables. Most computers use at least two types of storage.

In order to make the operation of the computer automatic, instructions must be fed to it by a 'control unit'; the instructions might have a form such as 'take the number with address p, multiply it by the number with address g and put the result in the location with address s; now go on with the next instruction'. The instructions themselves are coded as numbers, for convenience of handling, and stored in the same way as are the other numbers.

Preparing the instructions for a computer is called *programming*. If it were necessary to programme individually every operation to be carried out by the arithmetic unit, the speed of computation would be limited to the rate at which instructions could be supplied. The high speeds actually attained are due to the fact that it is possible to instruct the machine to repeat over and over again a series of operations, continuing either for a specified number of repetitions or until some particular number in store changes sign. A simple example of this is the method of dividing on a desk machine, where the divisor is repeatedly subtracted from the dividend until the remainder becomes negative (or, better, until it is less than the divisor). We now see how a computer can be designed to modify its programme in the light of the results it has itself derived: the great importance of the capacity for 'conditional transfer', as this is called, was fully appreciated by Babbage.

The only remaining items necessary in a digital computer are the means for extracting rapidly the results of its calculations, which are usually typed out automatically, and the means for feeding to it the data and instructions for the problem in hand. Here again, Babbage had realized that the punched cards used by Jacquard to control his loom provided just the thing needed. The programme and data for modern computers also are normally coded on punched tape or cards.

It will be clear by now how close is the parallel mentioned earlier between an automatic factory and an automatic computer. Facilities are required in each for transfer, control, storage, loading and unloading. Conditional transfer is analogous to the control adjustments resulting from measurement of the material in process; and inspection corresponds with the very elaborate precautions adopted in every computer continually and automatically to check its proper functioning.

Another computer designed in Britain, contemporary with EDSAC, is the ACE (Automatic Computing Engine), built at the National Physical Laboratory. Its speed and storage capacity of about 500 numbers are the same as those of EDSAC. For comparison, ENIAC's storage was limited to 20 numbers and *Whirlwind*, for example, stores about 1000, which is ample for general purposes. ACE was a pilot model for a rather larger successor called DEUCE (Digital Electronic Universal Calculating Engine), which was the computer used to predict the results of the 1955 General Election, during the early stages of the count. This machine, built by the English Electric Company, is commercially available at a cost in the region of £50,000. For its capacity, it is exceedingly compact and well designed, with a small high-speed store together with a much larger and rather slower one. Clearly, if a third machine is developed from this design, it should be a very high-speed computer and called TRICE!

A computer called LEO (Lyons Electronic Office), based on EDSAC, is in use for computing the pay-roll of a large catering firm, and a small general purpose computer made by the British Tabulating Machine Company, called HEC (Hollerith Electronic Computer), is available at a price in the region of £25,000. One is being used at the great oil refinery at Fawley for clerical duties and for computations in connexion with the plant operation.

The third major British computer to be built immediately after the war was installed at Manchester University. A dozen or so large machines based on this design have been supplied to all parts of the world by the firm of Ferranti, who have thus established themselves as the leading British manufacturer. When one of their computers was inaugurated recently in Rome by the Italian President, it played the Italian National Anthem and flashed 'Viva il Presidente Gronchi' on its monitor tubes! Ferranti now market a more compact computer called *Pegasus*, intended to be used for the same purposes as HEC. All the machines mentioned in the last three paragraphs take their information from punched cards and type their results.

Several causes, apart from increased demand and production, are likely to reduce considerably the size, cost and power consumption of commercial computers. One of these is the use of transistors and germanium diodes to replace valves: since they are no bigger than peas and their power consumption is only a small fraction of a valve's, the problem of cooling the computer will thus be eased and the total power consumption diminished. It is also hoped that the reliability of these new computing elements will ultimately greatly exceed that of valves. As a computer naturally contains many identical circuits, which are often repeated, it lends itself naturally to manufacture from printed or card circuits which can be assembled automatically, are very compact and can readily be replaced in the event of failure: this should reduce both the initial cost and that of maintenance.

The type and amount of storage capacity needed in a computer depend upon the sort of work it is to do. In accountancy a very large low speed store, possibly on tape or punched cards, is needed, together with a smaller high speed store for computing. Where there is proportionately less data and more processing, as in much mathematical work, a larger high speed store may be needed. ENIAC was used to compute the value of π to 2000 figures, and for this sort of job only a very small store is needed.

One reason why it may be useful to make a computer, like *Whirlwind*, to operate at very high speed is that it can work in 'real time'. This means that data can be taken from a real object (such as the motions of an aircraft in flight) and fed directly into the computer, to obtain information available in no

PLATE VII

AUTOMATIC ASSEMBLY MACHINE

This machine, called Autofab, can insert up to twenty-four assorted electronic components into printed-circuit plates, at a rate of twenty assemblies per minute. The drums, which contain the units ready for insertion, are replenished regularly by the operator and the plates are moved along the track beneath them.

PLATE VIII

LARGE TRANSFER MACHINE

This U-shaped transfer machine has an overall length of 162 ft., and a total length along the line of 350 ft.: it carries out 688 operations on V-8 engine cylinder blocks. The operator can be seen by the loading station in the foreground on the left. The automatic operations include 555 machining processes and 133 inspections; the production rate is 100 per hour, and 104 blocks are in the machine simultaneously: two of them can be seen to the left of the operator and two more in the final stations. The machine, which has eighty-seven stations in all, is divided into five sections with storage facilities for banking the blocks between successive sections.

other way. DEUCE was also operating in real time as it produced the election forecasts, which would have been useless if one had had to wait until the next day for them.

Analogue machines and simulators are inherently more suitable for problems in real time—they can even be made to go faster than real time, for the purposes of prediction—but their accuracy is generally much less than that of a digital machine: at the election, the men with their slide-rules had little difficulty in keeping abreast of DEUCE, for example. Speed of operation is thus one of the justifications for constructing an analogue machine like *Tridac*.

Whether the computer operation is analogue or digital, the main difficulty is invariably that of programming it—giving it its working instructions. This is an arduous and highly skilled job that only a very limited number of people can do —here is the chief hindrance today to the use of these computers. There are two parts to the problem of programming: the first is to put the problem into a form suitable for automatic computation—defining the physics of the problem— and the second is to code this into a form that the particular machine will accept. One of the major troubles at present is that, although the first part of the programming is of general application, the second—and more tedious—part is special to the particular computer used and valueless for any other machine. It is not even possible to make any other than the most minor modifications to a machine without invalidating all existing programmes which have been worked out for it. Simplifying the coding, or making it automatic, is the most immediate problem before the designers of computers, and plans for both are now in hand.

Let us now consider briefly how computers, or the techniques worked out for electronic computing, are going to be used in future automatic plant. They will have two main functions: performing clerical duties and controlling machines. We have already seen in chapter II how punched cards can be made to control a machine, as in Jacquard's loom. The tape can contain full instructions as to the actions to be taken by all parts of the machine, and so cause it to perform just as if a human operator were in control. Negative feedback is introduced to ensure that the instructions are being followed accurately and to make any small corrections that may be necessary. The use of tape control eliminates the slow and costly business of 'setting-up' the machine before it can operate.

One way of preparing the tape is the 'play-back' system, first developed in America (by the General Electric Company). A skilled operator makes the first piece on a special machine which records his actions; exactly the same series of movements can then be repeated much more rapidly, as there is no need to pause for thought. Alternatively, the control tape may be punched by an electronic computer, or the tape might be cut out altogether by making the computer give electrical instructions directly to the machines. This idea has led to the interesting suggestion that it might be feasible to telegraph spare parts to any place in the world, where there is a machine that could act upon these instructions.

Digital computers may soon be used in chemical plant, too. In order to obtain the greatest possible yield of the required chemicals in an oil refinery, the temperatures and pressures at which the various parts of the plant are automatically held ought to be adjusted continually, as the chemical composition of the crude oil supply varies (which it does, even when it comes from the same oilfield). This adjustment demands what is called 'end-point control'. The final product is analysed and the controllers are adjusted accordingly. To do this automatically, one will need continuously operating analytical equipment and also an electronic computer, capable of accepting information from the analyser and, in the light of this, feeding out the right corrective impulses to the various controllers.

Design of the master controller for a chemical plant presents great difficulties. The problem is to decide what information should be built into it. It seems likely that the solution will be for it to work on a 'trial-and-error' basis; making trial adjustments and then storing information about their results. The computer could, alternatively, work by rapidly assessing the results of each possible course of action before venturing upon it. It will, of course, be necessary for the computer to be capable either of detecting and rectifying its own faults, or at least of giving warning of them, and taking any precautionary steps immediately needed.

The use of a digital computer may also make possible the control of new chemical reactions which occur so fast that current methods of control would be inadequate. In the future, the computer will be as essential to the operation of chemical plant as a conventional controller is now. A master controller may also be used to control all the machines and processes in a complete production line, as this would be much more efficient than using many separate controllers, with each one at work for only a small part of the time. A controller such as this would also take over various clerical duties and operate on the information it gathers in such a way as to produce the facts upon which management decisions could be based.

Computers may also enable us to improve the performance of automatic controls, when they are applied to a development now being investigated at Cambridge University. This is the use of what may be called 'secondary feedback'. The 'primary feedback' is the monitoring signal which closes the control system loop; one can also measure some other quantity (such as the average value of error) to be used as a criterion of performance. This ' secondary ' measurement is then fed back, and made to effect an automatic adjustment of the sensitivity (or some other factor) so as to obtain the best possible performance.

In American industry the number of clerical and minor managerial personnel is nearly equal to that of direct factory labour. In so far as the work they do is of a purely routine nature it can be taken over by automatic computers and, with about a hundred of them now installed commercially, industry in the United States is already well on the road to full automation in the office. It seems certain that the routine clerical worker will finally disappear long before his opposite number in the workshop vanishes. It is not possible to describe the many American commercial products by name and it would be invidious to select a few for special mention; but their size, cost and capacity are for the most part comparable with the equivalent British designs we have considered.

Let us conclude this chapter by noting the words of a leading authority on the application of computers in industry, the Chief Organizing Accountant of the National Coal Board, D. W. Hooper. He said recently: 'The electronic computer is the finest tool in the field of management that has yet been devised; let us not hesitate to use it, but let us at the same time make sure we do not misuse it.'

CHAPTER VI

PRESENT IMPACT AND
FUTURE PROSPECTS

*Official of the Ford Motor Company to the President
of the C.I.O.:* How are you going to collect Union
dues from these machines?
Walter Reuther: How are *you* going to get them to
buy Fords?

We are not on the threshold of an age of fully automatic production. We are moving that way, but
slowly.　　　　　　PROFESSOR B. R. WILLIAMS
University College of North Staffordshire

An electronic and automatic age is with us. It is
not, I am quite certain, still many years in the
future.　　　　　　　　　　　　D. W. HOOPER
*National Coal Board and British
Institute of Management*

WE saw, in the first chapter, that the use of automatic control today is only a continuation of a
process that has been going on for centuries; but
it has become much more rapid in the last few
decades. Automatic regulators are found in almost
every phase of modern life, from the thermostat
on an oven to the automatic pilot in an aircraft;
they control anything from the voltage of the
electricity supply to the movements of guns and
searchlights; and it is certain that much modern
chemical plant could not be operated at all without
its automatic controls. We have also seen that
automatic control is only one of the technical

78

aspects of fully automatic production, though in some respects it is the most important.

We are now ready to consider the effect that the concepts of the control engineer are having on other branches of science, and to discuss some of the developments which are likely to occur in the next few years, particularly as the use of electronic computers becomes more widespread. The increase in automation will introduce some new problems for management and intensify old ones; the severity of these problems and of the social stresses resulting from automation will depend largely upon the extent to which industrial production can be made automatic and the speed with which this happens. These also are questions to which we must now attempt to find an answer.

When we were concerned with the behaviour of control systems in chapter III, we found that some of their most characteristic properties are accounted for by their having negative feedback. When a feedback system has time lags or multiple loops, we saw that a change in its sensitivity alone may be sufficient to cause oscillation. Now, it has long been observed that processes in fields other than that of automatic control exhibit this typical oscillatory, or cyclic, behaviour; and the suggestion that the concept of feedback could provide a unifying basis for work in many branches of science originated with Leibniz; he was the philosopher later lampooned by Voltaire as Dr Pangloss, who contended that 'everything is for the best in this best of all possible worlds'. Leibniz and Newton shared the honour of being the first foreign French Academicians; but, unlike his great contemporary, Leibniz died a pauper.

To denote the effect of feedback on political science the French natural and political philosopher, André Ampère, coined for it a century later the name 'Cybernétique', adapted from the Greek word for 'helmsman', to which we also owe our own word 'governor'; much more recently the feedback concept has been popularized, as 'Cybernetics', by Norbert Wiener, who has himself contributed to the mathematical theory and is deeply conscious of the social implications of automation. As there is hardly a phase of natural science that owes nothing to feedback, we can do no more now than consider a few examples that illustrate its widespread influence.

Many physiological processes depend upon the existence of feedback. The heart beats at its natural frequency for much the same reason as a clock does; emotional disturbance releases drugs into the bloodstream which change the properties in the feedback loop and cause the heart to beat faster. This is typical of many other automatic mechanisms that increase the chance of survival, such as the control of reflex actions—particularly interesting here are the movements, co-ordination and focusing of the eyes—also the control of oxygen intake and of fat storage in the tissues. Very important to survival is the control of blood pressure: when this drops, on account of loss from a wound, the peripheral vessels contract and numerous other mechanisms are called into play to reduce the loss and retain pressure in the body —for example, a powerful sensation of thirst induces the victim to replace the lost fluid by drinking.

It is necessary for the healthy functioning of the body that the blood should contain rather exact proportions of its various constituents: sugar, water, salts, calcium and proteins, for example; and the amount of each of these is automatically controlled. But the most important control of all perhaps, for warm-blooded creatures, is that which assures the constancy within narrow limits of body temperature. It is also one about which much is known: the temperature-sensitive element is a region at the base of the brain and when it is stimulated by an excessive temperature sweating is caused, so inducing loss of heat by evaporation; and the blood vessels near the skin are dilated, to assist the same process. When the body temperature is too low, on the other hand, the surface vessels are contracted to reduce the loss of heat, and shivering is initiated as a method of generating more heat.

We can stand upright on account of the reflexes developed in childhood in our muscular system. The state of tension in our muscles is continually being measured by sense organs embedded within them, and impulses from these organs are used to maintain this state unchanged, until overriding signals are received from the brain. It is particularly interesting that some of our sense organs are so constructed that the impulses they emit can actually anticipate the stimulus to some extent, and so greatly improve the quality of control. Disease or intoxication, on the other hand, may increase the sensitivity or time lags of these natural controls and so interfere with their proper functioning.

Physiologists use the term *homeostasis* to denote these various self-regulatory mechanisms in the body. Since those of them which are best understood are known to depend upon feedback, a reasonable and fruitful hypothesis is to assume that all of them are caused by feedback action. The hypothesis of natural selection is itself an example of feedback on a grand scale.

The study of the interaction of living things with their environment has also been assisted by feedback theory. When two animal populations exist side by side, with one preying upon the other, a rhythmic fluctuation in their numbers would be predicted under certain conditions, for this is the same situation as arises in a multiple loop control system. Just such a fluctuation in the numbers of rabbits and lynxes in Canada is well known as the 'Fur cycle'. What happens is roughly this: as the number of lynxes increases they eat more rabbits which accordingly decline in number, until a stage is reached at which there are so few rabbits left that the lynxes begin to die of starvation and so decrease in number; the rabbits thereupon multiply until they provide enough food for the lynxes to multiply again in their turn. And so the cycle is repeated indefinitely, if there is no external disturbance.

Something similar has been observed to occur with two species of fish and in this case it has also been found that the cyclic variation in numbers is inhibited by intensive fishery, just as would be predicted by theory. Such situations were analysed a quarter of a century ago by the French mathematician, Volterra, in his book *La lutte pour la vie*; but more recently Rashevsky, working in Chicago, has greatly extended this work, and has examined the sociological implications in his book *The Mathematical Theory of Human Relations*. This study of group behaviour, as it is often called, is an extremely complex and fascinating branch of sociology.

Some psychologists also are looking to control theory, to help them in their study of the functioning of the human brain. It seems possible that for many purposes the brain may operate by continually assessing the probable results of action (perhaps ten times a second) and then making appropriate adjustments. Such behaviour would imply that the brain has a huge capacity for storage of information—one might call it memory—and also the ability to do simple computations—which one may prefer to call prediction. Apparently the neural mechanism needed to accomplish both these tasks does indeed exist in the brain.

Many phenomena of engineering science can also be attributed to the presence of feedback. It has been shown by the author that this applies, for example, to the collapse of a long thin strut, and to the oscillation of an aeroplane wing, known as flutter. It is, perhaps, in economic theory, however, that the concepts of control can make— and, to some extent, are making—their greatest contribution. It is well known, for instance, that the price of pigs tends to fluctuate regularly, completing one cycle in a period of about four years; and it seems reasonable to expect that the so-called Business or Trade Cycles, of which this is typical, may be accounted for in much the same way as the Fur cycle, though the problem is very much more complicated.

Lord Keynes proposed a model of an industrial economy, which consisted in the main, he suggested, of two closed loops having as a common quantity the general level of economic activity (which is closely related to the standard of living). These two loops are the capital goods loop and the consumer goods loop; it can be shown that with such a system, a small decrease in the flow round one loop could be exaggerated by the other into a much greater drop in the standard of living.

When it is remembered that many of the dependencies in these loops are not linear, it will be appreciated that the analysis of even this simple model of the economic system presents formidable theoretical problems. The chief importance of this and later, more elaborate, models seems to lie in the constant reminder they afford us, that the economic system is both non-linear and multiple-loop. We discovered in the third chapter what unexpected behaviour such systems can exhibit, and it is wise to bear in mind that so little a thing as fixing the price of one commodity is capable of setting in motion a chain of events which could throw the whole economy into a state of instability.

As a particular instance of a radical disturbance of the industrial scene, we might consider, in a general way, what could be the economic effects of the present influx of automatic machinery and controls. Several arguments advanced in chapter IV indicated that replacing men by machines must increase the tension of the economic system; that is, it increases the sensitivity of some of the closed loops in the economy, which is just the sort of situation that can lead to instability.

Nevertheless, in spite of the risks involved, the economic advantages to be gained from automation are so great that there will surely be a vast extension in its use. As we have seen, the first industries in which it has been applied are those making relatively long runs of a product that changes little and for which there is an assured or even expanding market—canned foods, motor cars and plastics, for instance. Other articles which may well be made and assembled automatically in the near future are ball bearings, bicycles, typewriters, cameras, radio and television sets, clocks, ovens and refrigerators, bricks, fertilizers, soap and textiles.

As for the techniques of control, it is clear that developments already on the drawing board will be brought to fruition. We shall have automatic dialling of trunk and international telephone calls: the manual exchange will become completely obsolete. We shall have automatically controlled trains: already a driverless monorail train, running at speeds up to nearly 200 miles an hour, has been tested in Germany. We also discussed in the last chapter some of the ways in which electronic computers are likely to be used for machine and plant control, and to assist management with those problems arising from the generally greater pace which can be expected in all productive processes.

We have already discussed the management problems that may be introduced by the functional inflexibility of some automatic installations: another problem arises because the most appropriate way for a machine to set about a task frequently differs from the method suitable for the men it replaces. Though a mechanical robot makes a vivid symbol for the cartoonist, we are most unlikely ever to see 'machines like men, walking'. It would be quite uneconomic to construct a mechanism as mechanically versatile as a man, and then set it to work at the very simple tasks which would be all it would be capable of performing. Men are not machines, and it would be just as inefficient to attempt to make a machine like a man as it is to make a man do a machine's work.

This means that it may well be advisable to redesign plant to get the best out of automatic production. One may even go a stage further, and assert that often the product itself must be modified, to make the most effective use of modern methods of production. There is nothing new in this, of course, and one of the responsibilities of management is to convince the customer that such redesign is desirable. Already, advertisements assure us, for instance, of the superior elegance of streamlined shapes, because these are the forms which can be produced most easily by modern techniques of production. In order to achieve automatic assembly, many of our most familiar articles will have to be radically redesigned in the next few years.

Disposing of the flood of goods produced by automatic methods may be another considerable problem; the potential market is far from saturated, but the difficulty is with the distribution of purchasing power. Perhaps it is true, as Henry Ford remarked, that if consumption is to keep pace with production, then the working class must become the leisure class. It has been suggested that customers will refuse to buy the identical and standard goods produced by automatic processes; this may well be, but surely automatic production need not lead to uniformity: just as

Jacquard's loom could weave an infinite variety of designs, so could modern methods of control lead to greater diversity, provided the consumer elects to pay a relatively small amount extra for variety. It may well be that the first country to develop this approach will steal a march on its competitors abroad.

When installing and running an automatic line, it is most important to plan ahead to ensure smooth operation of the plant; maintenance must always be by anticipation, never by default. It is also necessary to be able to rely on a continuous supply of raw materials and a steady market for the product. Computers, employed on clerical, inventory and similar duties, will assist management to achieve these ends and because both computers and tape controlled machines can be very versatile, they may provide the answer we need to the inflexibility of much current automatic equipment.

But automation clearly raises as many problems as it solves, and we have seen that the most severe of these will be managerial and social rather than technical. One way of helping industrialists to meet their difficulties is by the introduction of novel methods of training for future executives, who must grasp clearly the potentialities of the new computers and acquire the fresh attitude of mind demanded by latest techniques. There is appearing, in addition, a new type of technologist —the control engineer—and special courses of instruction must be provided for the large numbers of such men that will be needed. This is a real challenge to our engineering schools and universities, as the training must cut across many separately established disciplines—a tendency to be welcomed as it will react against the current trend towards ever greater specialization.

Since proportionately fewer men will work in the fully automatic plants of the future, they will not have to be situated near large centres of population. Nor need they be near natural sources of power, since electric power—probably generated from nuclear energy—can be transmitted to them. The decentralization thus made possible may, in the end, be as radical in its effects on the distribution of population as was the Industrial Revolution. It will also materially affect the pattern of industrialization in undeveloped territories, which will surely evolve along lines totally different from those followed in the West.

But although great social changes are to be anticipated in the next twenty-five years, we should not expect anything more extensive than we have witnessed in the last twenty-five, for in spite of what some cartoonists may have led us to fear, not a single expert believes that any *sudden* change is likely to occur in our social or economic life, on account of these new techniques. For example, Lord Halsbury has said: 'I can see no prospects of the type of short-term social revolution which disrupts a society . . . the revolution is in the technological approach to production.'

What are the principal reasons for predicting that the changes ahead will be gradual rather than abrupt? The chief one is that it will take a considerable time to design and install the new plant that is needed for automatic operation, because current plant, as we have seen, is often unsuitable. This delay will be increased by the impossibility of meeting immediately the huge demand which will arise for electronic control equipment, though printed circuit techniques will certainly help with producing those units which can be standardized. Another deterrent to rapid change will be the immense capital cost of installing new plant in concerns like the steel industry.

The second main reason for supposing a sudden shift to automatic production throughout industry to be most unlikely is the problem of education. We must reduce the current shortage of technologists and skilled technicians, particularly specialists in electronics, and it is also essential to bring home to the management of some firms the great possibilities for automatic control in almost every field. As the President of the Society of Instrument Technology has remarked recently, the rate of progress will depend upon the closeness of collaboration between maker, user and research establishment.

Thirdly, we have already seen that automatic production is at present only applicable to some sections of industry, though it has been argued that some of the recent advances in control technique may shortly open the door to a far more widespread use of automatic production, even where the runs are quite short.

To summarize, it seems that the arguments presented in this book lead us to conclude that three widely held views appear to be in error. The first is that automatic devices can be made to do anything: we have seen that, even if economic considerations are disregarded, theory indicates that there are very definite limitations to what can be achieved. The second misconception is that the further use of automatic devices can make no radical difference to the economy, since it is only the extension of an existing process; with a complex feedback system such as this, however, no such conclusion can be drawn.

Thirdly, it has been held by Karl Marx and his followers that what he called 'technological unemployment' is an inevitable consequence of such advances as the introduction of automatic control. John Stuart Mill, on the other hand, reached the conclusion that while such progress may even benefit labour—and usually, in fact, does so—this is not necessarily always the case. We have seen that this is still the only position which can reasonably be maintained.

It might even be claimed that, just as the first industrial revolution was caused by the machines which replaced man's muscles, so may automatic controls, which replace his brains, be bringing about what Wiener has called the second industrial revolution. In wider fields of knowledge, moreover, what the feedback theory of Leibniz is doing to elucidate observed cyclic phenomena may be compared with what Newton's Law of Gravitation did for Kepler's observations on the planetary motions.

It appears, however, that we need not be alarmed at the social changes that will follow this great increase in automatic production. This optimistic view was strongly endorsed by Woollard when he spoke recently on the topic *Machines in the Service of Man*. He said then that 'Automation is not a device with which to outlaw, displace or dispense with man. . . . It is a means for increasing man's stature and extending his ability to produce in greater volume with less physical effort or mental strain.'

Whether we agree with this or not, there can be no doubt that in a highly competitive world the standard of living of any country, such as Britain, largely dependent on its exports, can only be maintained by a steady increase in its total productivity. It is most unlikely to retain or expand its export trade unless these more efficient methods of manufacture are used wherever possible. This point is stressed in another remark of Lord Halsbury: '. . . automation', he said, referring to Britain, '. . . is essential to the continued maintenance of our position in the world. . . . We cannot afford not to exploit it as intensively as we can.'

FURTHER READING

Although there have been plenty of contributions to the technical press and a good many popular articles on Automation, few books on the topic are yet available. But two of these are worth careful study: one, by Frank G. Woollard, is *The Principles of Mass and Flow Production* (Iliffe, 1954); and the other is by John Diebold and entitled *Automation: the Advent of the Automatic Factory* (Macmillan, 1952). Mr Woollard was one of the pioneers of automatic production in the motor industry and Mr Diebold led a group at the Harvard Business School which investigated the implications of automatic production. Woollard's book is the more technical and factual: his last two chapters are particularly relevant to our theme. Diebold's is more suitable for the lay reader. Both these books have bibliographies.

A more recent publication, concerned largely with labour aspects, is *The Challenge of Automation* (Public Affairs Press, 1955), based on material presented at a conference sponsored by the C.I.O. in Washington, D.C., in April 1955. Due to appear shortly is *Automation in Theory and Practice* (Blackwell, 1956), which is a reprint of a series of lectures given at Oxford University.

The Robot Era (Allen & Unwin, 1955) by P. E. Cleator, has just appeared. It is on a much more popular level than the books above but contains a lot of useful facts and some good illustrations. Novels with automation as their theme include Kurt Vonnegut's *Player Piano* (Macmillan, 1953), a satirical extrapolation of current American tendencies, penetrating but never bitter; and S. Mead's *The Big Ball of Wax* (Simon &

Schuster, 1954). These novelists' speculations are both technically plausible and socially disquieting.

There are many specialist texts on the theory and applications of automatic control. The following are introductory and not too mathematical: on process control, A. J. Young's *An Introduction to Process Control System Design* (Longmans, 1955); on position control, A. Porter's short monograph *An Introduction to Servomechanisms* (Methuen, 1950) and, with a general approach, R. H. Macmillan's *An Introduction to the Theory of Control* (Cambridge, 2nd impression 1955). The first and last of these have extensive bibliographies.

Books for the general reader on cognate subjects are W. Sluckin's *Minds and Machines* (Penguin Books, 1954), by a psychologist; Norbert Wiener's *The Human Use of Human Beings* (Houghton, Mifflin & Co., 1950), for a philosophic study; and edited by B. V. Bowden is *Faster than Thought* (Pitman, 1953), a very readable account of the evolution and operation of electronic computing machines. Wider in scope is D. R. Hartree's *Calculating Instruments and Machines* (Cambridge, 1950). Fascinating to read is W. Grey Walter's *The Living Brain* (Duckworth, 1953) and A. Tustin's *The Mechanism of Economic Systems* (Heinemann 1953), is worth noting. An interesting translation promised from the French of Pierre de Latil is *Artificial Thought* (Blackwell, 1956).

On various occasions, complete issues of periodicals have been devoted to branches of automation. The *Scientific American* for September 1952 was exclusively on 'Automatic Control'. *Metalworking Production* for 10 June 1955 was on 'Automation', *Machinery* for 10 June 1955 was devoted to 'Automation in the U.S.A.' and *Machine Shop Magazine* for June 1955 was on 'Electronics and Automation'. All these issues include articles by recognized authorities in the field. Of the many specialist journals devoted to our subject, two—*Automation* (Penton Pubs.) and *Control Engineering* (McGraw-Hill)—sometimes contain survey articles that are not too technical for the general reader.

The proceedings of the conferences held in Margate in June 1955 on 'The Automatic Factory' and in London in October 1955 on 'Automatic Control in Industry' have been published by the *Institution of Production Engineers* and in the *Proceedings* of the *Institution of Chemical Engineers*. They are among the most comprehensive and authoritative accounts of automation that are readily available and not too technical. Two sessions of the British Association have been of particular interest: one held in 1953 was on 'Cybernetics' and the other, in 1955, was devoted to 'Man and his Machines'; the papers presented on the first occasion are available in *The Advancement of Science* for March 1954; those at the second will appear in the same publication.

The broadsheet 'Towards the automatic factory' (*Planning*, 21, 380, 13 June 1955), published by P. E. P., is a good review of economic aspects. Useful articles have appeared in *The Times Science Review*, no. 15, Summer 1955 ('Continuous gauging' by E. I. Brimelow), in *Science News*, no. 37, Penguin Books, 1955 ('The automatic factory' by H. D. Turner) and in the *Financial Times*, 29 April 1955 ('Russia's Robot Factories'). Further articles, in *Mechanical Engineering*, are 'Controlling machine tools automatically' (June 1954, p. 487), 'Machine tool automation' (August 1955, p. 683) and 'Automation—its development in metalworking' (November 1955, p. 958). A short technical bibliography of 'The Literature of control engineering' was given by the author in an article in *Engineering* (3 June 1955, p. 687). Peter F. Drucker's 'The Promise of Automation' (*Harper's Magazine*, April 1955) is a remarkable example of clear thinking and exposition. Interesting also is 'The Pushbutton Factory' by Frank K. Shallenberger (*Smithsonian Institution Pub.* 4153, 1954). A paper on 'The Technology of Automation' by Ted. F. Silvey was considered by the U.S. House of Representatives on 15 June 1955, and appears in the *Congressional Record* for that day. A popular account of the first Russian piston factory is given by A. Erivansky in 'A Soviet Automatic Plant' (Foreign Languages Publishing House, Moscow, 1955). A penetrating pamphlet is 'Automation: a new dimension to old problems' (Public Affairs Press, 1955) by G. P. Schutz and G. B. Baldwin; and 'Automation—some social aspects', by H. de Bivort, from *International Labour Review* (December 1955), has recently been reprinted as a booklet.

On the historical aspects of automation, two papers read to the *Newcomen Society* are valuable. One, by H. G. Conway, is on 'The origins of mechanical servomechanisms' (January 1954) and the other is by A. R. J. Ramsey and entitled 'The Thermostat or Heat Governor: An outline of its history' (February 1946). Of interest also is 'The development of the Torpedo' (*Engineering*, 25 May 1945—15 March 1946) by Commander P. Bethell.

INDEX

97

For EU product safety concerns, contact us at Calle de José Abascal, 56–1°, 28003 Madrid, Spain or eugpsr@cambridge.org.

www.ingramcontent.com/pod-product-compliance
Ingram Content Group UK Ltd.
Pitfield, Milton Keynes, MK11 3LW, UK
UKHW010851090126
466816UK00011B/157